*For Joan*

# Contents

I

# Advice and Sandwiches

At lunchtime, Julie came out of Hodgeson & Co and clattered down the steps into the street. She kept her face down so no one could see how choked-up she was. Her numbers were good, her clients commended her, she'd worked extra hours so long she couldn't remember any other way — yet she'd been passed-over for promotion again. Julie blinked furiously. It was like being clouted on the side of the head with a brick.

Normally she would bounce right back. It was her inheritance from her dad — he had always been cheerful, always optimistic no matter what, to his very last breath. But that morning, Garry her boyfriend — no, her ex-boyfriend — had announced he was leaving her, barely a month after suggesting they move-in together. He said he'd met someone "really exciting". And he had that self-indulged look; he actually expected that she'd be delighted for him. And now this promotion thing. Swerving along the crowded sidewalk, Julie's face convulsed as she tried not to sob. In many aspects of her life she just didn't know which way was up. How could she bounce back when she didn't know which way was up?

She moved fast away from anywhere she might meet her colleagues, along the route she took to work each morning. A shop-space that had long been vacant was a buzz of activity. A voice was calling out, and people were straining their necks and laughing. A sign over the opening said *Advice and Sandwiches*, and a placard said *Queue here - Advice from €5 - Gourmet Sandwiches free*. The queue was the shape of a U, with a rope on little white poles to keep order. A burst of laughter rose from the crowd. It was a good-natured sound. Julie strode past. There was a cheer and clapping. She took a quick, grudging glance. People were saying 'Aww' like they were affected by something. A sandy-haired man wandered dazedly out, holding a wrapped sandwich. People in the queue turned to watch him go; they nodded knowingly to one another, until laughter from inside made them turn, crane their necks to see. Julie walked on, and the laughter faded behind. Street sounds closed around her. Everything was grey and worn and bare. Before the next intersection she wheeled around, retraced her steps. She joined the end of the queue.

She was too short to see what was happening. The queue was in three lines by now. A man's voice shouted "Next!" and a woman called out some sort of enquiry. Julie couldn't quite catch what happened, but things seemed to be funny. A woman left the shop looking embarrassed and amused; she scuttled off, gripping her sandwich in both hands. Next to leave was a younger woman whose sandwich hung in one hand. She moved slowly, wide-eyed, like she might bump into things.

'Next!'

Julie was no longer the last, and neither could she see over the people in front. A male voice called out: 'I hate my job. What'll I do?'

'Either make it so you like it, or else live cheaper, build up cash, then resign and look for a job you do like. €5. What you want to eat?'

'Aah... gimme a ham and cheese and tomato sandwich. No, a sub. A sub, please.'

'One ham, cheese, tomato sub. Look at that, ready already. How about that? Next!'

There was a low voice up the front.

'Can't hear you honey. A little louder.'

A woman's voice became shrill.

'My boyfriend's going to leave me and he only asked last month if I'd move in with him. What'm I gonna do?'

A murmur rippled around. Julie's skin froze. She waited with her mouth open.

'Honey, you might want to think – do you love him all over, or do you just like the way he looks? If you love him, then give it your best shot and keep doing it, but if it's everything or nothing with him then it ain't you he's thinking of. Start over and just really enjoy your life, and love will come. That's €5. What you want to eat?'

'Emm... oh anything...'

'Tuna lettuce and tomato pannini. Green wrapper. Next!'

Someone behind Julie reminded her to move along. She jumped to close the gap.

Many of the customers couldn't say what they wanted to eat. If they delayed, the man giving the advice ordered for them. Nobody objected.

'My wife's a bitch. What'm I goin' to do about it?'

This was a cocky voice, his friends egging him on.

'Go and ask her forgiveness for your own shortcomings. That's €8. What you wanna eat?'

'Hey, I thought it was five euros a sandwich!'

'Sandwiches are free – read the signs. Advice starts at five euros and goes up from there. You got the eight euros?'

'Aah…yeah, ok.' The voice was contrite now. There was knowing laughter, some jeering.

'What you want then?'

'Ham and cheese… aah… with mustard and pickle. On white.'

'Ham, cheese, mustard, pickle on white!'

'Can I get the green wrapper?'

'Next!'

Everyone laughed outright. Julie didn't get the joke - people were too close-packed to see. She began to feel nervous. At the first bend in the queue, she saw notices on the wall.

*Advice from €5. Free sandwich with all advice*
*No advice, no sandwich*
*Green wrappers share lunch between two*
*No loitering at counter*

A man wearing mirrored sunglasses was calling out the advice. He had three people behind him making the orders.

The line was moving at amazing speed. Julie began to fidget. A girl with frizzed hair and blue eye-shadow was waiting at the end of the counter. Julie thought it must be the girl whose boyfriend was leaving.

'Next!'

'Where's the best place to get a sandwich round here?'

A burst of laughter all around. It was a youngish man in a dark pinstriped suit. He laughed too.

'Right here Sir. Only honest advice and honest ingredients. €5. What you want?'

'How about a BLT sub?'

'One BLT sub, side order barbeque oyster and crispy cress pannini. Cut all that in small slices, green wrapper and give it to that girl there. You ok with that?' – he asked the girl with the blue eye-shadow. She nodded blankly. He turned to the pinstripe. 'You ok with that?'

The man looked taken aback, but he nodded. 'Sure, no problem.'

'Ok young lady, you take the green wrapper and you give him one slice at a time, I got the feeling his mouth is wider than his neck.' Everyone laughed. The pinstripe hung his head in mock shame. 'We don't want him to choke, right?'

There was a cheer as the pair left with their green wrapper. Everyone moved along. Julie felt exhilarated and nervous. Coming to the final line, she barely heard the questions. People were kneading their hands, hopping from foot to foot. Julie realised she was only three people from the end. She moved her elbows to cool herself. What would she ask? How quickly the remaining few were being served…

Julie was looking directly into the mirrored glasses. Her lip trembled. The man smiled.

'Is it about work?'

Julie nodded.

'Come right this way.'

He lifted the flap on the counter.

Julie mumbled: 'No no… that's not…'

But she tottered forward. What else could she do? The man took her handbag and put it in a cupboard.

'Wash your hands and make yourself a nice sandwich. Then get an apron and, well, start making the orders. Ok?'

Julie nodded slowly. She filled a roll from ingredients near her; roast beef and cold baked vegetables. The sandwich-maker at the end, a stout middle-aged man, slid tubs of mayonnaise and salad toward her. Julie smiled at him. She sat on a blue plastic crate. One of the other makers, a red-haired woman, gave Julie a cup of water and a smile in the easy stride of getting lettuce from a fridge.

Julie ate her roll like she was hovering over the scene. The sandwich-makers worked quickly but carefully. They spent time getting things right. Julie finished her roll and sat looking. The man with the reflective glasses gave longer advice depending on how busy the sandwich-makers were, although one time he gave advice for so long all the makers became idle.

Julie took a place at the worktop. She found disposable gloves, a paper hat and an apron. Next time a sandwich of the ingredients near her was called, she made it and handed it to the sunglasses man. After that first sandwich, time rolled like honey from a jar. She made sandwich after sandwich in a daze.

'Next!'

'What's the best sandwich for someone who loves cats?'

'A shared sandwich. That's €5. What you want?'

'Oh? Emm… chicken, some salt, light mayo.'

'On plain?'

'Plain.'

'Chicken, some salt, light mayo on plain, green wrapper. Next!'

'Yeah, umm… how do I get to share lunch with that girl?'

'Bring flowers tomorrow, something small for her cats, and don't expect a thing in return. Maybe the day after you might get the green. €5, what'd you like?'

'Umm… Chicken roll. Just chicken. Is there smoked chicken?'

'One smoked chicken roll…'

'Emm… excuse me? Excuse me?'

'Yes Ma'am.'

'I don't mind, I mean… could that man get the green wrapper today?'

'Ma'am, you can ask anyone to join you any time you want, you don't have to ask me…'

'Yeah but… you know…'

'Ok, make that smoked chicken roll in the green wrapper. Thank you. Next!'

A young man with a knapsack was next. His hair was shiny, curly at the sides. He looked awed.

'I'd like … I don't know…'

'What advice d'you want?'

'Advice? Emm… what could I do that's kinda worthwhile, I guess?'

Everyone laughed. The young man looked around, bemused.

'Well, Sir, so happens we have a vacancy coming up right here.'

Julie and the other makers looked up. There were only four spaces. They looked at one another to see who might be leaving. The counter-hatch was lifted and the young man came in just like Julie had done. He seemed happily bewildered. The man in the mirrored glasses smiled at the makers.

'Do any of you guys have sunglasses?'

The makers quickly shook their heads. Except Julie. She had sunglasses in her handbag. The man in the mirrored glasses looked directly at her. Julie felt a thickening in her throat. She had been enjoying herself so much. How long had she been there? She had been floating. Her dad used to say all things come to an end, there had to be an end so that there could be a beginning. She looked at the new young man. He had beautiful eyes. She turned to the man in the mirrored glasses.

'I have sunglasses.'

'Ok! So after this fella has himself a sandwich, if you give him your apron, that'd be great.'

Julie made sandwiches until the young man was finished eating. He stood, awkward and self-conscious, while she put her apron over his head. She told him how the making was organized, and it was astonishing how much she could tell him. Everyone beamed at them. His eyes were green and brown, and they were honest. When he was ready to make his first sandwich, she cupped his face in her hands and she kissed him. Julie barely heard the commotion. The man in the mirrored glasses said: 'Ok Miss, let's see those glasses.'

Julie got her handbag. She took out the sunglasses. With sunglasses on, nobody would be able to see her eyes, and she was glad of that. It was a good idea. She said to the man in the mirrored glasses: 'I'd like to thank you. Really. It was great.'

'No, thank *you*, Miss. We appreciate what you're doing.'

He reached out and she shook his hand. Then he lifted the hatch and he walked away past the queue. She stood staring after him, her hand still extended.

'Hey Miss! What's the best stock on NASDAQ for a quick punt?'

Two young men in suits at the front of the queue, jostling each other. They found the question hilarious. Julie gaped at them. She said to the guy at the front: 'Do you own the place you live in?'

'Me? Uuh...no?'

'Then you can't afford a punt on the NASDAQ. That's €8. What do you want to eat?'

# Keeper Hill

Timmy Ryan drove up the forestry track until a black and yellow barrier blocked the way. He parked in a small clearing, and stood slowly out of the car, stretching his back, flexing his hips to relieve the stiffness. The air felt cool under the dark branches, birdsong was shrill and clear. A shaft of sunlight shone blindingly through the dense spruce canopy, down to the ferns and heather that bordered the track. Droplets glittered on cobwebs, wild flowers glowed like jewels. Timmy took a deep breath. How he had missed these walks.

He laced-up his boots, lifted his knapsack from the car, and checked that he had water, a banana and a bar of chocolate. It was August, and this was his first and last walk of summer. Normally he walked once a month during the summer, with his friends Ger Hickey and Laz McElligott, but this year every walk since April had been cancelled due to heavy rain. It was the third summer in a row that rainfall had broken all records. Today, neither Ger nor Laz were available. It was the last week of their summer break, and both had been pressed into foreign holidays by wives despairing of ever getting any sun in Ireland. Normally, he wouldn't have been here today either, because Limerick were in the Munster football final, and although they had not won this title in over a century, this was now their fifth time reaching the final in fifteen years. Timmy had rarely missed a match even in the bad years, but today he had been unable to go. The match was in Cork, and Emily, his daughter, had been receiving an award for basketball in the University of Limerick. He couldn't have missed that, and afterwards he had driven Emily and his wife, Anne-Marie, to the train station in Limerick Junction, because Anne-Marie was taking Emily shopping in Dublin, to the Dundrum Centre. This was a special treat for both women, because Emily would be leaving soon to go to university, and Anne-Marie was missing her already. Of course, at the train station they had tried one last time to make him come too, and he tried once again to make them understand that he really did not feel in the slightest bit left out, in fact he was torn between watching the match in a pub, or going up Keeper Hill for a walk, on the only decent day there had been in months.

'Why don't you go for the walk and listen to it on your phone?' Emily had said. He had stared at her, not actually knowing that such a thing was possible. She laughed, took his phone and programmed-in the radio station, showed him how to use it.

Out of pure stubbornness, he decided not to bring even a light coat in his bag. Keeper Hill was the highest in the region, but still it was less than seven hundred metres, not steep or rocky, and it had a track all the way which most people used. It would be an easy walk listening to the match. He looked at his watch – ten minutes to throw-in. He locked the car and moved up the track past the barrier, untangling the earpiece wire as he went. From this point, he would need to go cross-country for a bit, to join the normal hiking trail two-thirds the way up. He cut off at the first switchback on the track, and made his way directly uphill. The sun warmed the side of his face. He could feel moisture rising from the ground, making the air dense, almost palpable.

He stopped at a weathered rock outcrop, took off his pullover and stuffed it into the knapsack. Looked at his watch again. Time to get the radio working. All he had to do was plug in the earpiece and turn it on. But... what the hell? It was some sort of talk show, an argument about an earthquake in Greenland, for Christ's sake. Timmy glared at the phone display: it seemed to be the correct station, right enough. He looked at his watch – throw-in was in just two minutes. Surely they couldn't have a different programme right up to the beginning of the match? Could there be a change of schedule? Now that he was psyched up for it, he couldn't miss this match. He turned off the telephone, stamped angrily up the slope. Maybe he should go back down right now. Drive to a pub and watch it there. His boots made a crunching sound in the heather. Tiny pale moths flittered out of his way. He would wait until the match was actually starting before he tuned-in again, then he would know for sure. A bumblebee droned past – a sudden, pleasant sound. That bee almost made him smile. Last year, when he was walking Mount Brandon with Ger and Laz, a bumblebee just like that had flown between them, and Ger had commented that bees were becoming extinct due to agricultural monoculture.

'Monoculture is necessary,' Laz had retorted, 'otherwise humanity would starve. Which would you rather, grandchildren or bees?'

Ger said: 'There's more people on the planet this century than in all history combined. Just think about that. There'll be ten billion by the time we die. There was one billion when my father was born.'

'Ok, so reincarnation can't be true, then, can it? But still you don't want intensive farming?'

'I don't want intensive people. Who the hell's advising the governments? Economists like you! So you can have expanding markets, for another forty years anyway.'

'Weren't you bitching when the economy tanked and we had the wage cut? You wanted expansion then, didn't you?'

They had argued all the way up to the top of Mount Brandon, and all the way down the Saints Walk to the minibus. Timmy smiled. Ger and Laz were both lecturers in the University of Limerick: Ger a geographer, Laz an economist; and they were opposed on nearly every subject, so walks were entertaining.

Timmy tried the radio again.

'...collapse on that scale is not possible...'

'Oh for heavens sake!' he muttered. 'Come on!'

'You only say that because no modelling has been done.'

'But gentlemen, was it an earthquake, yes or no?'

'There are no active faults in Greenland, so it cannot be an earthquake. But that's academic at this point! What we've to deal with right now is the repercussion, and what I'm...'

'May I point out, Dr. Farrell, that it is completely alarmist...'

'No! No! There is absolutely no doubt. We've no warning system in the Atlantic, but a tsunami will definitely be underway! The only doubt is the magnitude. And we're already forty-five minutes into this. What we have to do at once...'

'I think Dr. Farrell is jumping to conclusions. Not every earthquake results in a tsunami, and this ...'

'For the last time — this is not an earthquake! This... it looks like a widespread collapse, a catastrophic collapse in fact, of the Greenland icesheet. What happened to Reykjavik was a result, not a cause. If you just look at the data...'

'With due respect, you cannot get 9.5 on the Richter scale from an avalanche.'

'Professor Hausemann, we are not talking avalanche, we are talking an ice *sheet, half the size of Europe...'*

*'But gentlemen – gentlemen! – what would cause such a thing?'*

*'Look, water's been pouring into the lower parts of that icesheet for years, we know that. We know it's not freezing anymore, so there's a hydraulic head under that ice. So that... that ice-sheet is like a giant hovercraft. And the sheer amount of rain...'*

*'But Dr. Farrell, we have measured a seismic shock...'*

*'I know that! But continuing for twenty-five minutes? At 9.5? What kind of earthquake is that? I'm looking at the readout while I'm talking to you, and there's no epicentre... what are we saying?– this's totally academic at this stage, we're wasting time! We need to alert people, that's why I'm on the phone! At the speed these things... Jesus, these things travel a thousand kilometres an hour, d'you realize that? This thing's more than halfway here already! Do you understand what I'm...'*

Timmy removed the earphones. If this was one of those War of the Worlds-type panic dramas, it wasn't funny – a Munster football final was on. He was plodding through an area of sphagnum moss that sucked at his boots, and he had to move sideways on squelchy tussocks to firmer ground. A sweet algal smell rose from the ground he had disturbed. Flies and midges gloried in the sunshine. On the open slope, the day was windless but cool, the air clear. He looked back to where the conifers peeped above the curve of the slope, to the foothills beyond with their undulating hedgerows and tiny fields. It was postcard pretty. In the hazy distance he could see the Clare Hills, and, at their foot, a greyness that was the city of Limerick. He imagined the bustle down there, where he had been less than two hours ago. An idea occurred to him – Radio Limerick – surely they would have the match. He would have to tune it himself (that would impress Emily). He looked around for a place to sit, moved diagonally to an area of lichen-covered scree and found a reasonably flat surface to sit on. He started scrolling through phone menus, and eventually found how to tune the radio. He inserted the earpiece, scrolled through snippets of music, and suddenly he was in the boisterous excitement of the match. Limerick had conceded two goals – Timmy felt himself slump – but at least they were clawing their way back, point by point. Limerick won the kick out.

'Come on! Jesus' sake, come on will ye!'

On the deserted hillside, birds looped away from his glazed eye. To his right, a trio of red deer hinds wafted down to the conifers. The Limerick attack broke down. Kerry worked the ball upfield and scored a point.

'Oh Jesus! Come on, will ye? Come on!'

Midges began to itch in his hair. He stood abruptly, scratching his hair and arms, and started again up the gentle slope. He watched his boots squeeze the water from blossom-pink sphagnum, his head loud with commentary and crowd's roar. Limerick scored a point, and Timmy made a fist. He found himself striding along, grimacing or twisting his body with each turn of the match. He did not slacken until he reached a tumbledown cairn between a concrete trig cone and a television mast. This was the top, but he could not sit while the match was on. He decided to continue onward, maybe to a vantage point southward over Tipperary, where you could see to the hills of Waterford and Cork in the distance. If he walked until half-time, he should have exactly the right distance back for the end of the match.

Brittle hilltop heather crunched underfoot, the panorama was blade-sharp in the cool air. In the earphones, roars of elation or disappointment rose and fell. On the dot of half-time Limerick put the ball wide, advertising cut in, and Timmy yanked out the earphones in frustration. Imagine, two goals down after eight minutes! He ground his teeth. He would have liked to debate Limerick's man-marking defence and high-ball attack – if only Ger was here (Laz would be useless; being from Kilkenny, he didn't know a thing about football). Timmy turned off the phone, partly in frustration, partly in case the battery might drain. He pictured himself sprinting down the hill to listen to the final minutes on the car radio. He would do it. No question.

He came to a stop on a slight mound. It was good to stop at some sort of feature. The lowlands greyed into the distance. Fields and hedgerows, houses and hillocks, churches and villages. This was what Ger Hickey would call a panorama of industrial agriculture, habitat destruction and embattled ecosystems. Laz, however, would see a landscape in the economic likeness of man, constantly changing with the demands of human need and technology. What Timmy himself saw were the football strongholds of Templemore and Tipperary Town, and Thurles further off, a mountain in terms of childhood memories, where as a boy he had attended matches with his father and uncles. All of this lay low and still in the August afternoon sunshine. Only the merest drone of distant traffic. Or was it blood-flow in his ears?

He finished the chocolate, gathered the wrapper and the banana skin into the mesh pocket of the knapsack. He turned on the radio, put the earpieces in his ears. What? Oh no. He hadn't saved the tuning – it had reverted to national radio.

'...planes have all been diverted.'

'But what does this mean? For us in Ireland, I mean. Who can we ask? Do we have...? No? Dr Farrell? Is Dr. Farrell back on the line? Yes! Dr. Farrell, the Institute of Advanced Studies... you heard what they're saying in Air Traffic Control? Can you tell us what this means? I mean specifically for this country. For us.'

'Yes... I mean no, no I didn't hear what they said. But I... the research vessel, Celtic Voyager – we'd asked them to assess the swell, and they've reported an upwelling of... of tens of metres... in the open ocean. That... that could translate ... I don't know... on land, that could be hundreds of metres – I hope there's some error here, I really do. This is without precedent. No one – no one – has any idea what we're looking at. We're onto the USGS in the Pacific right now, trying to formulate some appropriate response... but we're... we're rechecking the signals... I don't know–'

'Dr. Farrell, if the dreadful, dreadful happenings in Reykjavik were caused not by earthquake, but a tsunami from Greenland, could such a tsunami still be moving eastward? Could it be moving toward Ireland?'

'What the hell?' Timmy said. 'What in the name of God is this?'

He yanked out the earpieces and glared at the phone. Around him, all was peaceful. He scrolled along the frequency bar to the local radio station.

'...back the lead once more! What an answer from the Kingdom County! Kennedy finally gets into the game and what a way to make his entrance! Now let's see what Limerick are made of! Let's see now, whether the Treaty County ...'

Timmy loosed a breath in relief. They couldn't possibly be covering a football match if there was some disaster afoot, could they? Anyway, he would listen to the match, then find out about this earthquake business – it was only another twenty-five minutes or so. But maybe he should phone Anne-Marie first. Or Emily, who might actually hear the phone. Just to see. But they were in Dundrum Shopping Centre, they wouldn't have heard any news. He tried Emily. The phone said "Network busy, please try again later."

Timmy stood motionless in the silence of the hillside. A yellow and tan bird cheeped its annoyance. It fluttered short bursts in a wide circle around him. A butterfly made its languid, erratic way across the heather. The only sound was a distant jet. He tried the phone again. Waited. Tried again. Perhaps this place was too remote. Or maybe that TV tower interfered with phone signals. He would try further down the slope. He moved briskly back the way he had come. Switched on the radio.

'...*much slower on land, but still it would be in the order of a hundred and fifty kilometres an hour...*'

'*A hundred and fifty kilometres an hour? Across land?*'

'*Yes, and that's a wave of water hundreds of metres high, and it...*'

The radio became silent. Timmy stared at the phone display. He found the correct menu, scrolled again to the station with the football. Nothing. He peered closely to see if the phone battery had faded; no, that wasn't the problem. He tried phoning Anne-Marie, then Emily. Network busy.

Timmy stopped walking. Took the earpiece from his ear. The day was as calm and still as any he could remember. It reminded him of his childhood. Not a person in sight. He looked toward the horizon. But at this location he was facing northeast. He had a sudden, overwhelming need to look west. He began to contour the hillside in an ungainly run. He stumbled over rocks, through clumps of heather. Grazed his shins. Banged his knee and had to limp. Gradually, the low hills of Clare came into view. They were between him and the sea. He had to stop. Sat on a tussock of heather, panting. Took the phone out again. His hands trembled slightly. Network busy. No radio. He decided to text Emily. And Anne-Marie. Just in case. She rarely looked at her phone when she was shopping. He waited. Wetness seeped through the seat of his trousers. He kept his eyes on the Clare hills. All his life, nature programmes had been warning of calamities. Carbon dioxide, methane: all these vague concepts. Nothing ever came of it. But Ger Hickey said that in the natural world, things got worse slowly, slowly, until things were really bad, and then they got much worse suddenly. Like going broke, Laz said. Or dying, Timmy had said, remembering the death of his mother. He missed his friends now, but not in the way he missed Anne-Marie and Emily. If Anne-Marie and Emily were with him he wouldn't be concerned at all; he would calm them and they would be together. Gazing across the broad valley, it seemed that the marks of mankind were barely more than scratches on the landscape. Roads and quarries were only a few metres deep, you could hardly see them from here. But with this God-like view, he could easily imagine that valley filled with a swirling torrent of water. Nothing could survive such a situation. A wave that size would compress all of history into a moment of destruction, like sandcastles on a beach.

He sat on the tussock, panting, gazing across the valley. It was too late to change anything now. How could you change things? In just a few minutes, everything would be exactly as it always was, or would be forever changed. Should he run down the hill? Climb the TV mast? He imagined people all across the country, sitting rigidly, waiting – like he was sitting here. There was nothing else to do. On the horizon, the hills of Clare seemed dark and uncertain. He could not see them properly. Was it a line of hills, or a wall of water? In just a few minutes, he would know.

# Creeping

I think I saw youngsters in the garden this evening, sneaking near the gate. The whole place is covered in bushes and trees and that's the only place you can see anything. I can't see much anyway. But I saw them.

I'd been meaning to try for milk and bread. Usually I go in the early morning, when no one is around. But often it gets too late. The pavements get busy, and I have to wait another day. This evening I really have to try, cos I've nothing left for days now. Milk was delivered here before Gracie died, when all this was flowers and lawns and paths. Now branches rub the windows, leaves block the gutters. A few short years and the trees are in. It's like they lie in wait, watching for you to become weak. They send in feelers like those youngsters do, testing the boundaries. One day, you look up – they're all around you.

Gracie wouldn't recognize the neighborhood now. It used to be… not well off, but it was good here. Everyone knew everyone. Everyone was the same. Then, after Gracie – they built those apartments. Looking right down on us practically. They broke Gracie's memory with that one stroke. Then three houses where the O'Malleys used to live next door. I stayed inside, to let things settle. To let them stop changing things. And they did, eventually, when they couldn't build anything else. They sent me a letter – would I consider selling the house? Gracie's house. That was the last letter delivered here. I went straight down – *furious!* – to the post office. No more letters please. No, not one. Ok, a collection box, I'll get them when I'm want them.

The pavements have become full of people. I don't know any of them. Young impatient couples. Prams and buggies. Toddlers, children squealing. Gloomy youths hanging around, making comments and sniggering. I thanked God then for the bushes. I didn't have to see those louts sitting on Gracie's wall. I could hear them though. Out on the street – bellowing and screeching. Especially in the evenings. On school holidays, it goes on day and night – it's like the jungle. I dread calm days now. I pray for rain – it keeps them indoors. Any moment they might put the windows in on top of me. I've only one room left, rot is closing-in like the bushes. But I must have food – milk at least. I can't carry on. But I can't go out. If only I could get milk delivered.

They're in the garden this very minute. I can hear them. Rustlings. The crack of a twig. Low voices. Somewhere in there. It is too much.

Behind a paint-peeled door, I find Gracie's rake. It's the size and weight I need, because my left hand doesn't close anymore. I grasp the rake, tremble as I push along the path. My neck is very stiff. I can't see forward when I stoop under branches, can only see my slippers inching along. Near the gate, the voices are in the laurel – low, miserable voices. My heart may burst. I have to wait a few minutes, until the pulse is slower. I say to myself – this is it. The time has come. I raise the rake, ease myself into the laurel.

There is a space, a cave under branches, with a sheet of plastic like a tent. I cannot see all of it, but I see a boy and girl, maybe ten years old; he is squatting, she is kneeling on cardboard. They are arguing sadly, putting something in a bowl. The boy has his back to me. The girl's face is wet, or is it just the light? I grip the rake more tightly, inching closer. In the instant that she sees me, I am thinking of a den I had once near a railway line, and I almost falter. But it's dog-eat-dog. I go in swinging the rake. The boy is too quick, but I get the girl on the knee. There is a flurry – screaming – a bursting through bushes. How many others there are, I do not know. I turn to make my escape.

By the time I reach the house, wheezing has taken all my energy. I lie against the half-stuck door until I get my breath. I judder the door in, try to close it. But it is swollen, sagged. It does not close anymore, so the chain is my lock. Leaves have gathered in the hall, ivy is snaking its way in. At least the chain still makes it. Thank God for the chain.

In the room, rubbish is piled against the walls. The only place you can still see the carpet is in front of the armchair. That smell of old milk cartons, the drone of the half-buried fridge, the crusted, dirt-shiny arms of my chair; these are my comforts. A brown stain has been spreading across the ceiling. It was halfway when I noticed it a long time ago, and a small piece of the bulge has fallen. I have a radio too, but I cannot play it:

it fills me with the worries of other people. Also, I cannot hear sounds if the radio is on, and I need to know if they're coming. But the rake: I have dropped the rake. What betook me to use it? There could be reprisals, I see now. So I sit, cowled in my blankets, with the breadknife ready. Nothing comes. Is that a mercy or not? I no longer have strength. I want to get it over with. I would like them to come while my pulse is racing.

Evening has deepened. Darkness falls. Out in the street, roars and shrieks echo against the walls, the blare of loud cars and motorbikes is angry, ever-changing. I can't know who might be in the garden. The windows are useless because of the bushes. I dread those windows. I've dreamed of sheets of wood, of corrugated iron, so that when they come they must at least come by the door. But corrugated iron is as impossible as anything else. So I wait, for the twinkling sodium-yellow shapes on the wall to become grey with dawn. Then I can go out, onto the pavement, at least for milk.

When the room is bright, I boil the kettle and get the teabag. There is plenty of colour in it still. Enough for two days maybe. But I know I must not wait till the last. Not like last time. This time even the sugar is gone. I've pulled apart the old packets I can find, of sugar and biscuits. I've shaken them, even licked them. Lack of food is like a siren muffled by my skin. Black tea keeps me going, but it doesn't sustain me. I take so long at the tea that when I get to the front door, the shriekings of children echo from the street. I have missed it again. I must wait until tomorrow.

A day comes. Everything seems very bright in the room. When I stand up, the room around me is like a faded photographic negative. I wait a long time before I can risk moving. But I need something. The teabag would take too long, and I have not another day to lose. There is no sound of youngsters, so I release the door-chain and start along the path. Even under the bushes everything seems unusually bright. Yet I cannot distinguish twigs or leaves. Near the gate, fear brings my shuffle to a stop. What if they are waiting? What if their burly parents are here? I wish I had the rake. I push into the laurel, fearful, to get the rake.

How beautifully still everything is under the bushes. All dry and peaceful – no movement, no noise. The plastic sheet is still there. The rake is there, fallen at the side. In the place where the girl kneeled on the cardboard, there is a white tub with some water, half-knocked over, and a blue plastic bowl with brown stuff in it. Bluebottles buzz from this brown stuff, and settle again with their darting runs. It is food. My stomach bubbles. The tongue sticks to my lips. This will be degrading, I know, but I am a long way from the shop and nothing is for certain. I will have to get on my knees. There is a box I can lean on, under the plastic sheet. In the box is newspaper, and something dark is on the newspaper. An animal. I pull the box out to see. It is a puppy, smooth and tan coloured, lying on its side. It has too much skin and it is very still. But there is movement at its eyes. White things, like crumbs of bread. Oh, they are maggots. The eyes are gone. I nudge the body, and a ripple moves along its fur. Its paws curl minutely. It is not dead.

I look away, into the stillness of the bushes, and see there an image of myself, lying on my side, waiting for the flies to realize that I cannot even blink. And later, when my tongue cracks, the magpies and crows and rats will come. There is no pity under the bushes. I kneel for a long time.

Eventually the shouts of children begin, as every day. I only vaguely hear them, because I am thinking now that I must take up the rake and finish-off this unfortunate creature. I lift the body from the box. Its tiny sagged movements, the twisting of the maggots, are excruciating to me. I lay the poor thing on the cardboard, and maneuver backward to where the bushes overhead will allow one good swing with the rake. I raise it, bring it down as hard as I can. Raise it again, to be certain.

There is a snap of a twig, and I freeze – it is the boy, creeping. He is almost in the clearing before he sees me, with the rake and the shivering body of the pup. His face has barely time to stretch before he flounders and blusters in the bushes and is gone.

It hits me, what he must feel. That I killed his pup. That is what he will believe. In his head, he will make war against me. But in his soul he must surely know – I know he will know – that he has simply been too long away. Like I was, for that single weekend, when I went to a conference of steam trains in England. I was away four days. When I came back she was on her face in the hall. Her arms were perfectly at her sides. Everything was absolutely perfect, but she was cold and stiff.

The boy has dropped a plastic water bottle – it is still wobbling – and a box of something. That, I am sure, is food, and I think maybe, maybe… if I can just… But it gets very bright under the laurel. Far too bright. I know in my heart that it is too bright altogether.

Something is moving against the light. The boy. He is in the bush again, in a glistening plastic coat. There is the smell of earth and wet leaves, the all-around tip-tap of raindrops. He bends low, cautiously, like he is peeping into a burrow, to see into my face. Water streams from his hood. His black wellingtons creep toward me. I am a skin filled with broken glass. The boy edges forward, bends wide-eyed, leans. In my hand, the handle of the rake has become darkly wet. His soft pink fingers curl around it, and slowly he pulls Gracie's rake from my grasp.

Oh yes, I think – a bang on the head would be like a comforting quilt. If only he has the strength. But he jumps backward, casts the rake aside, and without taking his eyes from me, he gets out one of those phones, and begins haltingly to speak.

# Prizes

Best patterned tights.
Best... curl of hair at the ear.
Best shiny leggings.
Best... graceful turkey neck.
Best alien walk.
'Spare coppers, sir...? Ah sure, God bless anyway.'
Best candy lipstick.
'Spare coppers? Oh God bless. God bless. Thanks very much, sir.'
Best high heels.
Best... best... post-colonial tweed – ha!
Best tight leather jacket.
Best... patent handbag... oooh, no no no – best perfume-wake.
'Spare coppers, sir...? Ah, God bless anyway.'
Best... half-shade glasses... no, no – best inanimate expression.
'Spare coppers, sir...? Oh, God bless. God bless you.'
Best high heels – no, had that already – best... best... aam... best hair grip! – phew.
Best legs in jeans. At the front anyway.
Best curvy calf muscle – oh, contender for calves of the week, definitely.
'Spare coppers, sir? Ah sure, God bless anyway.'
Best inch of slip – ha ha!
Best... ummm... youthful overdressing?
'Spare coppers, sir? Oh God bless, thanks. Thanks very much.'
Best ... eyelashes.
Best – oh Lord God... best imitation of Margaret! But surely...? Exact tilt of head, swing of hips! Hair! Even her hand coming down, as if Cathy, as if... little Cathy...

'Paul, is that guy following us? He is!'
'Hey, what's your problem there, fella?'
'Oh sorry sir. Sorry miss, I was only…'
'Go on, back off there now.'
'Misunderstanding, sir. Just… misunderstanding.'
'Well, would you stop staring at my wife!'
'But miss… have you ever … been in an accident? Or have you ever–'
'Paul, he's some kinda nut!'
'Lookit, buddy, you wanna clear off outa here.'
'But sir, if I could just –'
'One more step, buddy, God help me I'll flatten you, cripple or not! D'ye hear me?'
'Ok, ok, sorry sir. Mistake. Just a mistake. Sorry.'
'And don't be following us, you got that? Or I'm getting the guards!'
'Sorry sir, sorry, apologies.'
'Flippin' weirdo…'

Best… shoe.
Best… foot.
Best umbrella drip.
Best… best…

'Whoo! You ok there, man? You ok?'
'Is he alright? Here, get him up.'
'All together on three, right? One two three – hup!'
'Ok… ok. Fine. Thank… thanks. Thanks very much all. I'm… I'm alive. Thanks very much.'
'Here, sit against that, man. There you go.'
'And here now, will you get yourself some food with this, will you? None of that drink or drugs. Will you do that?'
'God bless you, ladies. God bless… Thanks.'
Aah.

Best… waist in woolly jumper.
Best peep of tummy.
Best… cheerful nose.
Best… Oh best lover's glance! Fabulous! Glance of the week contender! Winner in fact. You should celebrate. Enjoy it. And good luck in the finals.
'Any coppers sir? … Nooo problem. God bless.'

# Tulips

Sharon groaned at her reflection in the mirror. Another overcast morning, early June, and in the grey light her eye sockets appeared even deeper than they already were. Ray, her husband, had glued this mirror to the bare concrete blocks; it was slightly crooked and a little too high, but better than nothing. This corner of their bedroom was to have been the en-suite, but it had not been plastered or wired, nor the partition built, and only the toilet and hand-basin had been installed. Sharon put her fingertips to her cheekbone, angled her face to see the lines gathering around her eyes. A little makeup would help, but she didn't have anything left that contained a tint. Anyway, she had to go downstairs and look after the girls, get them fed and dressed so that her mother, Granny Trish, could collect them later on, and she and Ray could go to their Saturday jobs. Sharon lifted her handbag, plodded out through the door-gap in the wall, and down the uncovered wooden stairs.

The smell of concrete was always worst in the mornings. She put her hand over her mouth, gulping softly. For more than a year she'd had vague indigestion, but for the past couple of months a pain in her stomach had woken her almost every night. Last night she'd hardly slept at all. Lying there waiting for dawn, she had come to the formless realization that she could no longer carry on. She had to see a doctor. It was the very last thing they needed.

So she told Ray, when he got up that morning, that she was going to a hairdresser before she went to the call-centre. Why did she say such a crazy thing? They'd been cutting each other's hair for the last two years. They each worked two, sometimes three jobs, because the practice where she worked as a solicitor could now only offer three days a week, and this combined with Ray's teacher salary could not pay the mortgage and the loan. They needed every single cent they could get. Ray drove a taxi three nights and at weekends, and she worked three nights as a bartender. On Saturdays she worked in a call-centre in Clonmel, seventy minutes drive away.

But what did Ray say? Great – he said – I'm glad you're doing something for yourself.

How could he believe she'd spend money on herself?

She stopped on the stairs, wiping her cheeks, sniffing back tears.

Nausea welled in her stomach. She rummaged in her handbag for antacid tablets. In the bare echoey hall, the crinkling of foil and plastic sounded incredibly loud; she hunched over the packet in case the kids might hear. Popped two tablets, took a deep breath, and clipped briskly toward the kitchen.

The sight of it made her droop. Boxes for cupboards. Crappy cooker, crappy crappy fridge. Bare joists for a ceiling. A forty-bottle sommelier cabinet, never plugged in but unsellable, that they used as a worktop.

Ciara and Aoife sat dawdling at their porridge.

'C'mon,' Sharon called. 'Hurry up, let's go. Daddy's going to be working here today.'

'But can we help, Mummy, can we?' Ciara with those big dark eyes.

'Not this time, Ciara. He's going to be making cuts in the walls, and it'll be really dirty and dusty. You've to—'

Little Aoife squeaked: 'But Mummy, can we watch the telly so? Can we? We've that Minecraft game that—'

'No. But you can take it to Granny Trish's. Later.'

'Pleeeease, Mummy!'

Sharon had to lean on the table, clench her teeth. 'Look, can't you just play outside till Granny comes? The whole place'll be covered in dust, I told you. '

She heard Ray opening the back door, stamping the mud from his boots. He spent half an hour every morning at the vegetable patch, so that if every credit card failed and every cheque bounced, they would at least have something to eat each day. It gave them an illusion of security. That and the apple trees his Uncle Eamon had given them when they bought the site. 'Long term business,' Uncle Eamon said, winking. 'Like houses and marriage.' Eight trees with huge roots wrapped in sacks. They had planted them one November afternoon three years ago, their first involvement with the site. Ray did the heavy digging, Sharon and the girls arguing about which tree should go where, like it was the most important job in the world. Whenever Sharon looked at those trees, she remembered

Ray's silhouette in the slanting sun, the girls with their toy spades and watering cans, bossing each other and patting the soil.

Her private vision of living in the country had been of tulips – masses and masses of tulips. It was something she'd carried since childhood, from a film perhaps, or a photograph. Ray had planted two hundred bulbs for her between the driveway and the hedgerow, and now each springtime those tulips made a glorious blaze of yellow and pink and red, even though they were shaded by bushes and trees. At the time, she had pleaded with Ray to cut the hedgerow down, but even back then there had been more important things to do. Now, time for gardening was out of the question. Still, last Sunday when she had burst into tears after dropping an egg on the floor, and he said 'That's ok, not to worry,' she told him he didn't understand one *fucking* thing about her, he hadn't even cut the hedgerow down. Normally he was such a steady guy, but that day he stamped off into the rain, borrowed a dreadful-looking chainsaw, and spent the whole afternoon cutting that hedgerow until it was time for his taxi shift.

Now Sharon watched him toe-off his muddy wellingtons, plod along the passage toward the kitchen. How thin his face had become – she hadn't noticed before. What would happen to them if he got sick? She turned quickly away, hurrying the girls, trying to keep calm. It was 3.30am when he got home from his taxi shift last night. Today was Saturday; probably it would be four am, maybe five.

'D'you need more breakfast?' she asked shakily.

He puffed his cheeks. 'No. Better get started.'

He had borrowed some big electrical thing called a con-saw. A friend of his brother had made chalk marks on the concrete walls where wires were to go, and Ray was to cut either side of these lines with the con-saw, then chisel-out in between so that wires could be fastened-in and eventually plastered over. It would be the first stage of getting proper wiring, instead of extension cables like they'd been using since they moved in. Ray could manage rough manual jobs like that, jobs that involved brute strength, but he was awkward and clumsy, his attempts at plastering and carpentry had been a waste of time and money.

'Come on!' Sharon's called to the girls. 'Coats! Boots!'

Ray sighed. 'You go on, I'll manage them. And enjoy yourself.'

Sharon tensed.

'What did you say?'

He looked up blearily, confused.

'How the hell,' she hissed, 'could I enjoy myself?'

He made a wide-eyed shrug, let his eyes fall.

Would he even notice when she didn't come home with a hair-do? The effort of restraining herself made her tremble.

The whole thing was her fault. His job had been to look after the actual building project, her job to look after the legals and the money, and it was her idea to get a bridging loan, to not sell the old house until the new one was ready. Pay contracts up-front to get fifteen, even twenty per cent discount. She had organized the mortgages, conveyancing, all the agreements – it made perfect sense – maximize their money where it really counted, by paying in advance. It all seemed to fall apart in one day, but in fact little Aoife went from three years old to almost five while the kitchen contractor, the landscaper and finally the builder himself went bust. They had to sell their old house for barely half the outstanding mortgage, and move into this building site. But did Ray say anything? Did he chide her? Did he say one single, solitary thing about the whole house of cards? Not a word. Not one single, stupid word.

All he did was work harder. Week after week, evenings, weekends. Mixing concrete by hand. Wheel-barrowing sand, carrying blocks. Humping roof-tiles up ladders to save money. She couldn't do that kind of thing, she just couldn't do it. Last week men came in a van – grim-faced men – shouting for money owed by the builder. Ray had stood out there reasoning with them. He offered to remove the stairs, for God's sake. Sharon had been listening at the window, twisted with worry in case they took him up on it. Fists were shaken. He was prodded. Insulted. She heard herself being insulted. But Ray didn't lose his temper. Ray never lost his damn temper.

'Have you had your own breakfast?' he said softly.

She stared at the saucepan of porridge with its thick-skinned overflow – a rotten, sulphurous feeling bubbled within her. She turned quickly away, to see little Aoife eating nonchalantly, mouth open, hair stuck to her cheek; Ciara, spoon halted in mid air, eyes unblinking. Sharon's scalp tingled. She pressed a hand to her stomach, made dizzily for the door.

'You do the kids,' she called, and slammed the door so hard the windows shook.

The air outside cooled her temples. She sank into the car seat, weak with relief that she'd kept her stomach contents down. But where was she going? Not the bloody hairdresser, that was for sure. She ground her teeth, hauled on the steering wheel. The car swerved around, down the rubble driveway. To the left, a swath of cut-down branches lay as they had fallen across the faded tulips. Her tulips. She bounced the car out onto the road. Hedges zipped past. Damn those branches. She drove faster, but the memory of the branches clouded her vision. The road blurred in front of her. Maybe he'd knocked those branches onto the tulips deliberately. She pulled-in to a gateway, jabbed the phone. It took him a long time to answer, and all she could hear was muffled panting, a powerful electric motor winding down.

'Listen,' she said, 'could you...! Hello? Hello!'

The snap of a dust-mask being removed. 'Hello? Yes... Yes!'

'Could you do just one thing – go out and pull those stupid branches off my tulips.'

'Look, I told you, remember? I'll do it first chance. But I've only got today–'

'You said that a week ago!'

'I've been working on the house.'

'I know that! Don't you think I know that? It wouldn't take you ten minutes to clear those branches.'

'Half of them are thorn bushes, you've to cut off the thick bits and press the–'

'All I'm asking you is get them off my tulips. It wouldn't take five minutes!'

Somewhere in the background, Aoife squealed in demand. Ray groaned.

'The tulips are finished for the year.'

'The *leaves* have to have light.'

'They're getting loads of light.'

'Next thing the whole place'll be overgrown.'

'It took two days to do the bushes behind the house, remember?'

'Look, it wouldn't take you one minute–'

'It'd take at least half an hour, at least, just–'

'If I threw my rubbish over your precious apple trees, how'd *you* like it?'

'Eh...? What d'you mean – *my* apple trees?'

'Just cos they're *my* flowers, the only thing–'

'Wait a minute, what d'you mean–'

'So just because you don't give a damn! You don't give a single damn! They're my tulips, so you say *Fuck them!* Cos you blame me for all this!'

'What...?'

'That's your way of getting me! You don't give a *shit* about me now!'

'Look... you're being–'

'–me or my tulips, the only single thing I've ever asked you for!'

'I'll go and do it—'

'Ah don't bother! I'll do it myself!'

'I said I'll do it and—'

'Do whatever you like!'

His voice became thick: 'I will *clear*—'

'Precious bloody apple trees! Precious bloody children!' Sharon wept. 'You don't give a shit about me! Cos you blame me for all this mess!'

She waited hopefully, but there was only his breathing, the children yelling behind. She clicked the phone off.

Tears dripped onto her blouse. She gripped the steering wheel, stared through the windscreen at the sun-dappled valley, at roofs of old slate and the spire of Newport church. Why had she accused him of favouring the children over her? All she wanted was for him to shout at her, just once, to say out loud that she had ruined everything, and then they could argue, have it all out, and maybe start again. Not Ray, he was too proud, too bull-headed. But everyone has a breaking point. She pictured him striding furiously outside, ripping the cord of that chainsaw, it's awful banshee wail… then Ciara grabbing his arm – 'What's wrong, Daddy?' – and the chainsaw swinging around.

Sharon's fingers covered her mouth. No, that couldn't happen. She shook her head. Fumbled for the phone. Hit his number. The phone trembled at her ear. Long insistent ring tone: she held her breath. The call rang out. She hit his number again. The ringtone blared like a fire alarm. No answer.

'No no no…'.

She twisted the key in the ignition – it made a terrible noise – she ground and ground the starter, cursed it fervently, until the roaring engine made her realise it had been running all the time. She spun the car around – oh, how did she get so far from home? She wasn't supposed to come this way at all, she was halfway to the call-centre. She swerved around bends, bumped over margins. It seemed to take an eternity to get home. Finally, there was the gap, their gateway. There were the branches. But no Ray.

She skidded to a stop at the side of the house. The chainsaw was not under the wheelbarrow. No sound of that con-saw either. No sound of children. Sharon's knees wobbled. She tottered to the back door. It was hanging open. The plastic mat was skidded aside, and oh… there were drops and smears on the concrete floor, deep raspberry red. Sharon braced herself against the walls of the passage. Inched her way along. No sound of television. The kitchen came into view. Ray slumped on a chair, breathing heavily. Staring. Blood on his hands, all over his jeans. He raised his eyes to her, spread his hands in apology. But she rushed past, around the corner toward the living room. Burst open the door. Ciara and Aoife on the couch, heads bopping, earphone wires tangled in their hair, playing a game on TV. Ciara turned, moon-eyed, and Sharon made a creeping run, grabbed them both by the head, hugging them.

Little Aoife squealed: 'Mummy – oow! What're you doing! And why've we to stay in here?'

'Oh shut *up*!' Ciara yelled. She grabbed Sharon's blouse. 'Mummy, is everything ok? Mummy! Is everything ok?'

Sharon nodded jerkily, eyes closed. Abruptly her eyes popped open; what about Ray? She lifted her hands urgently, frantically, to silence the girls, stop their questions, so she could get back to the kitchen for Ray. Aoife turned back to the game, unconcerned, but Ciara would not let go of Sharon's blouse.

'Mummy, what's happening!'

'Shush... shush – everything's ok! Everything's–'

'But why are you crying! What's–'

'Please, Ciara, please... I have to go–'

'I'm coming with you!'

'No no, you have to stay–'

'But what's wrong! Why are you–'

Sharon hugged her fiercely, despairingly, then held her at arm's length, looked in her eyes.

'Ciara, listen to me! Listen! I have to go, and you have to stay here. Daddy needs a bit of help, and everything will be fine as long as I can go and help him. Ok? But it's... private. Ok?'

'But Mummy...!'

'No, you stay... Stay!'

Ray was still wide-eyed in the chair. She laid a hand on his shoulder.

'So,' he said thickly.

One thigh of his jeans was ripped to the knee in a pattern of sideways cuts. Twists of skin and denim were matted together.

Sharon gulped. 'Are... are you...?'

She kicked off her too-loose shoes, rushed off for antiseptic and cotton wool, sprinted for her repair box, to get the blunt, flaky-chrome scissors to cut away his jeans. She dabbed and swabbed gently at the wound, lifted his leg to take off his blood-filled boot. Then she saw the chipped ivory of his kneecap, and dropped everything, phoned 999 for an ambulance.

They waited, heads down.

'I'm sorry,' he said.

'No, I'm sorry. I just didn't see... I thought... the girls...'

Ray shook his head. 'They didn't see it yet.'

She shot a glance at him – the big innocence of him, he had no idea what she'd thought he had done. She sniffed deeply. Blinked back almost-happy tears.

'They'll get over it,' she said brightly. 'We all will. It'll heal.'

He stared at her, not seeming to understand.

She nodded: 'They can even come to the hospital maybe.'

She tried to sound cheerful, as cheerful as possible, wringing blood from his sock into the sink. She tried a smile. Outside the window, some pale discs on the grass caught her attention. But they weren't disks, they were the creamy-white centres of tree stumps, where the young apple trees had been. The trees were heaped in a ragged pile, the chainsaw lying in the skid-streaked grass.

Sharon's mother collected the girls, and Sharon stayed late in the hospital. In the greenish glow of a surrounding curtain, Ray sat hunched on a bed, Sharon upright on a chair. They each sat looking at their hands.

Ray said quietly: 'We could tell them the trees had a disease.'

Sharon nodded.

She said: 'We could get more trees.'

Ray's throat made a swallowing motion.

'If you think it's worth it,' he said finally.

Sharon nodded slowly, then more quickly. She raised her eyes tentatively, uncertain if it was ok yet to lay her hand on his.

# The Fly

After my usual breakfast of cereal and yogurt, I dawdle into the front room to start work. My job is supply chain management, and since broadband became instantaneous, my work clothes are T-shirt and pyjama bottoms. I can do practically anything without even going outside. Yesterday I had my lunch sent from England, ordered a footstool from China, and received a potted plant from Borneo, to add to my collection of carnivorous pitcher-plants from around the world. I think we should stop evolution right here, because we are at the apex, in our T-shirts and pyjamas.

I've just tended to my new plant, put the bell jar back over it and sat at the laptop, when I get this creepy little tickle on my tummy. I lift the T-shirt, and – yikes! – it's a little beetle or something. I try to brush it off but it's so *quick*, I can't seem to get it. I jump to my feet and flap my hands and shake like hell and – phew! – there it is on the table. It's only about the size of a split-pea, quite flat, and now I'm annoyed that I got in such a welter. I see that it's actually a fly, with wings tucked close to its body, sort of finely striped, safari style, like a renegade from a jungle. The cheek of it, sneaking into my apartment.

Naturally, I decide to kill it. So I slap it hard, and although I definitely get it, it speeds out from under my hand. I slap it again, and it's still not dead. The damn thing's as hard as grit. And so quick! I slap again – aaah! – it's on my hand! I flap and rub but it moves like hell. It's between the fingers, it's on the back – whichever way I turn my hand it's on the other side. And then it jumps – it's into my armpit.

I stumble backwards and the whole table is upset, the laptop crashes to the floor and I'm wrenching off the T-shirt and scrubbing like hell at that armpit. But where the hell is it? I get the tickle in the other armpit and I scrub there with the balled-up T-shirt – there it is! – it's onto my back. I whirl and slap and I see it dart under the band of my pyjamas and now I really, *really*, go absolutely nuts.

I get those bottoms off in a blur, rub the T-shirt furiously in the crack of my bum. Where is it? It's in my pubic hair. I do a hysterical run-on-the-spot, scrubbing and scratching and rasping with my nails. It does spidery little darts, like it's doing a dance. I mean… what's happening? This is *my body*!

I sprint to the bedroom, douse the little menace with talcum powder, spray it with deodorant. I empty the whole can on it. It takes no notice at all.

The shower! I bolt for the shower, start the water running. Maybe it's gone….? No – it darts across my tummy, and I am straight into that freezing shower.

But the fly scoots around my belly like it can dodge the drops. It seems to be enjoying it. It makes a million little darts; this way – stop, that way – stop. What the hell's it doing? I'm red from slapping at it. The shower gets scalding and I howl. The fly moves even faster.

Bubblebath! – it's my only hope. Thank God there's a bath here. I jump out and get the bath going, splurge the whole bottle of Radox into it. But the fly takes-off from my shoulder. It's the most awkward flyer you ever saw but it lands on the window, nips out the little opening at the top. Boy, do I slam that window closed. I check all the other windows, and finally relax. Thank heavens.

For a minute there I thought the world had gone mad. I mean, really, where did that thing come from in the first place? Why does it even exist? When you think about it, what is the *point* of that fly?

The next day, there is a hot little itch where I first saw that fly on my tummy. I take a good look. There's a small pink patch on the skin. Actually, it's a mass of tiny pink dots. I begin to feel itchy all over. Especially my crotch and my armpits.

A thought comes to me.

Are those little dots… eggs?

# King of the Pippins

I remember the original apple trees in the walled garden, the bark of the low branches silky-smooth because of us swinging on them, the higher branches crispy with blue-grey lichen. Because he was the eldest, Pierce would sometimes claim the biggest tree as his, or sometimes the tree with the biggest apples, or the sweetest apples, or the reddest. It all depended on which game we were playing. A tree with long yellow apples became the property of Dermot, because he was lanky and blond. Alison liked a tree that was low and easy to climb, and whose apples were soft and sharply sweet. Even our mother had a tree, though she had no part in our games. Hers was a small, delicate tree, with sweet, ruby red apples. It just seemed to match her character, so no one else ever claimed that one. Being the baby of the family, I always had the last choice, but anyway I had to choose one with low branches, otherwise one of the others had to come to help me climb it.

At different times, trees became male or female, stupid or clever, sporty or fat, brave or cowardly. Sometimes the trees were jungles (in which only very particular animals could live), mountains (snow-covered or volcanic), continents (with indigenous warring civilisations), houses (with rooms according to the number of branches you could negotiate), or even banks (Dermot's idea, with different shapes and sizes of currency, trading at varying exchange rates depending on ripeness). One time, the trees became castles, the apples were ammunition, and ferocious bombardments ensued, until Alison was hit in the eye, and a general scatter took place before she could bring that news to the kitchen.

In our family world, nothing seemed to change for a very long time, until it became an object of amazement that Pierce would soon be leaving to go to university. After that it hardly seemed any length before Dermot and Alison went too. Afterwards they moved further away to find work, and still I was in school. Our parents had become gradually more frail and the walled garden fell imperceptibly into disuse. By the time I was a late teenager, several of the ancient trees had already died, the remaining ones overgrown by briars and creepers, and upstart sycamore and ash, seeded from beyond the high wall.

If you asked me now, I couldn't say what happened to my own years of university, or the many years I spent working on oil rigs. I enjoyed the years on the rigs, more than I should have maybe, but all that remains of that period is a memory of getting back out to the rig one time, simmering over a row with my wife Colette, and receiving a phone call. I'd never had a personal call to the rig before, and immediately I feared for Colette.

It was Mum on the phone. I listened with a sense of unreality to her dainty telephone voice, asking how I was, what the weather was like, and how was Colette. I lied and told her Colette was fine, thinking Mum had probably heard about Colette's drinking and that we were having counselling and maybe that's what the call was about. Mum said: 'You know, your father fell off a ladder.' It was almost a normal thing to say, undramatic. Afterward, whenever I wonder about change-points in my life, I come back to that simple utterance. At the time, I asked calmly if he was ok, and she said yes, he had broken his collarbone, that was all, he was fine, and then I made jokes about it. She asked how the new apartment was working out, and I told her Fine, we'd spent a lot of time trying to get the bathroom right. I didn't tell her what Colette had done to the kitchen during one of her binges.

'What he was doing,' Mum said, 'was trying to fix the roof on the kitchen extension.'

'Really?' I said. 'But that's the best bit of roof on the house.'

'No, it's the worst bit now. Flat roofs don't last.'

I remember, at that point, having a sense of the world turning, of realizing my place under the sky. In my mind, I was still a child of an old unchanging place, but it wasn't so.

'Will you be coming down, next time you're off?'

There was a slight tremble in her voice, but I couldn't begin to explain to her why I just couldn't go down there for the foreseeable future, and she didn't tell me that rain was getting into the kitchen and they couldn't afford to do anything about it. We said our goodbyes awkwardly, as was the way in our family.

By that stage both Pierce and Alison had almost-grown kids. Dermot's children were younger, because he had married late, or maybe because he had a high-pressure job in Brussels. I was the only one of us still living in Ireland, if two weekends out of five could be called living.

A week after Mum's call, my sister Alison phoned in a panic, to tell me we had to organize care for Dad, and a visiting rota.

A couple of years after that, Dad had a serious stroke, and we all made the journey home. There was an overwhelming sense that it might be the last time we would be all together. Colette, of course, was the only spouse not to come. The hospital had let Dad out, but he had to stay in bed, with quiet visitors one at a time. It happened to be Halloween, and the kitchen was crowded, with Dermot's gangly French children tearing around in high excitement. Talk turned to the Halloween parties we used to have, but there were no apples anymore to play dunk-apple, or apple-on-a-string, or neck-pass. We talked about the old trees: the Fizzy Crabs, the Pudding apples, the Waspers, the Crow tree. A spark seemed to kindle in our mother's eyes, as she remembered many things that we had forgotten. Pierce and Alison listened to her, enthralled. Normally I would have loved to sit and listen, but I had troubles of my own. I needed to get outside, work-off my frustration on something inanimate. I said: 'How about clearing the walled garden? Anyone on?'

No one had any interest, never mind the energy. They bullied some to their children to help me, which lasted about ten minutes.

The walled garden was completely overgrown. Not even the tip of an apple tree was visible. I spent a good two hours thrashing about with a billhook, and made no impression at all.

Next spring, on one of my weekends to look after Mum and Dad, I hired old JP, who used to do odd-jobs for Dad. JP was handy with a chainsaw, and together we hacked our way into the regenerating forest that had taken over the walled garden. Dozens of young sycamore and ash trees were crowded in there, competing for light, as well as elder, blackthorn, and densely verdant lime trees that had started life as posts for chicken-runs and washing-lines. It is astonishing what nature can achieve in a couple of decades. The place was unrecognisable, and I was dismayed.

'Best thing you could do,' JP said, 'is get-in a big machine. Level the lot in wan go.'

There was some sense to that. The undergrowth tangle of briar, nettle, and creeper seemed impenetrable, the wild trees blocked out the sky, and weekends were brief. But I wanted to see if there were any apple trees left.

Eventually we found one. It was dead, but I knew from its weeping-willow shape that it was the Fizzy Crab. None of the trees had been really large, but this one was tiny compared to my recollection of it. It diminished my memories, but at least it gave some sense of where in this young forest the other trees might be. I gave JP the bad news: no big machine.

JP cut sycamore and ash trees into sections, and I dragged these to the woodshed. I watched him tamp-down a cluster of briars around a knotty old bole, and take his stance to cut.

'Hold on a second!' I shouted

He revved the chainsaw, leaned toward the base of the tree.

'JP! Hold on!'

I flung a twig at him. He looked up.

'Haah?'

'That's an apple tree.'

'Yerra, that's no apple tree.'

It was a gnarled-looking stump wreathed in creeper, but I remembered its shape, its impenetrable presence on the periphery of our games. Its fruit had always remained unripe until all other apples were gone. It had been what we called the Dad's tree.

'There's ne'er a leaf on that for years,' JP said.

I pulled a filigree of ivy from its bark, and there was a cascade of woodlice. JP was probably right.

'Well,' I said, 'we'll leave it anyway.

Next time I was home, we piled-up the cuttings for burning. JP did most of this work because Dad's situation was coming to its inevitable conclusion, as was my marriage, and all I was good for was slashing miserably at briars. JP lit a huge bonfire, and I leaned on the billhook, remembering the bonfires of May Eve and Halloween evenings long ago, with all of us working in the mysterious gloom of evening. The Dad's tree seemed suddenly significant. I went to take another look at it. But it was gone.

'Where's that old tree?' I shouted.

'Haah?'

'That old apple tree that was here.'

JP gazed around with his red, droopy eyes. 'There was no apple tree there.'

It started to rain. We heaped branches onto the fire until the downpour beat us, and I had no choice but to go indoors, and JP to go home.

Some weekend after the funeral, I thought I might as well burn the remaining portion of that bonfire, and there, upright in the centre, was the old apple tree. The fire had been built around it. I gazed at it awhile. Then, with undue fury, I scattered the unburnt cuttings away from it, left the scorched remains of the tree poking mournfully skyward.

By springtime, the walled garden had its first covering of grass, and two scraggly clumps of daffodils that had survived somehow. It was a cheerful sight. I helped Mum into the rocking chair, in the scullery that we'd converted into a garden room.

She said: 'That tree was always last to give up.'

'What tree?'

'The only tree left.'

She raised a quivering finger. The blackened old tree had a flicker of green on one part.

'It was good for tarte tatin,' she said.

Those words instantly evoked the long-ago battles in Malachy's field; Gerry and Frankie Twomey, and Noel Joyce, laying siege to us in the Toply Bushes, pelting us with blackberries or elderberries or whatever they had, while we collected Squeegee Whites. Then our war-cry would sound: "Tarte Tataaain!" And Pierce and Dermot would charge, whooping and yelling, chase the boys all the way to the Bottomly Bushes, me on my little legs coming along behind.

I went outside to look. Two tiny branches had green shoots.

From that time on, every time I came to visit Mum, I would go out and have a look. After three or four years, a cluster of three hard, acorn-sized apple buds appeared on one of the branches.

I don't remember if Mum ever made tarte tatin again, but on a beach some years after that, I was distracted from my book by the blood-freezing cry of my own little children, Saoirse and Brian, as they charged toward the waves.

"Tarte Tataaain!" they yelled.

Marge, the woman who looked after Mum, phoned me to say that a huge tree had fallen into the walled garden. When I got down there, I found that a beech tree from out the front had knocked part of the high wall and obliterated the old apple tree. Time to call JP again. He was very elderly by that stage, but nothing revived him like the fumes of a chainsaw.

He cut the beech branches to manageable lengths, and I pulled these aside so he could get at the main trunk. The old apple tree had snapped a few inches from the ground, but when we pulled the last of the beech off it, a thin branch sprung upright, with a few burnished-yellow leaves on it.

I whooped. 'Hey-hey! There's the apple tree you nearly killed twice.'

JP lifted his watery eyes, gazed with distain on the ragged sprig.

'Well,' he said, ''tis shagged now anyhow. You may cut it or lave it, 'tis all the one.'

'We'll leave it.'

Some years after Mum died, I moved back to the old house. There was nothing to keep me in Dublin. Saoirse and Brian had long since finished university and moved away. I was divorced, inevitably, and was retired a couple of years early with Parkinson's disease. While I was still able, I took on a low-budget renovation of the house and garden, and decided to replant a few apple trees. A local group called Seed Savers specialised in heritage varieties of all sorts of garden produce, and two nice young ladies came to give me advice. I showed them apples from the old tree, and explained that these stored well, had a smoky flavour and were good for making tarte tatin. They promised to try to identify the tree, and took some cuttings to propagate.

A card arrived saying that the apples were King of the Pippins, an ancient variety, originally from France. They had spliced the cuttings and these were doing well. Which was good, because like myself, the old tree was in poor condition. The bole was riddled with woodworm and half of it had no bark at all. Shelves of fungi grew from the bare wood, and the branches were so crusted with moss and ferns and lichen that the whole thing looked like bedrock. Every year I expected it to die.

But it lasted. And lasted. Finally I thought it was going to outlast me.

Some summer after that, there came a peculiar gale from the southeast. I looked out the following day to see King of the Pippins lying on its side, its rotten roots up in the air. I left it as it lay. The leaves withered, and most of the unripe apples fell off. Strangely, a few remained, hoping perhaps to swell and ripen one last time.

Saoirse and Brian were good enough to come on alternate weekends to help me manage. They wanted to have the tree sawn up, because the grass could not be cut with it lying there. But I wanted to let it lie. All that autumn and the following summer, I let it lie. The place will go to weed, Brian said. Yes, I nodded, it's nature sending its emissaries, to see if we're still around. The tree was soon covered by goosegrass and climbing-weed. Brian was living in Geneva at that stage, and when he came home, he spent most of his time out in the garden. It was the first time I'd known him to take an interest in gardening, but one look at my withered and shaking limbs, and I couldn't blame him.

Now, finally, I have to let them cut up the tree. The man they get to do it is JP's son, Dano. He's a bull of a man, like his father was, and he doesn't wear earmuffs either. It's Saoirse's turn to be here this weekend, which isn't easy for her, because she is pregnant with her first child. She puts a chair out for me, at a safe distance, so that I can watch the proceedings, and she stays foostering around, in case I might try to stand up and get involved.

The chainsaw roars. It is indecent how quickly Dano goes through that tree. I feel every cut in its limbs. Soon there is only a line of rust-coloured rounds, a dusting of saw-chips and a tangle of twigs. Dano leans-in to cut a twisted spike that remains connected to the ground.

'Wait!' Saoirse holds up her hand.

'Haah?'

'Hold on a minute.'

She peers at something, and comes to fetch me.

'What…' I say.

'Come and look.'

She helps me across to the tree.

From a shard of trunk, a branch as thin as a pencil is heading for the heavens. There are three green leaves on it.

Saoirse looks at me. Her hand is on her bump, and in her eyes I see that she wants me to hang on, just a few more months. But I have no say. It takes two saintly women, day after day, to keep me going. Lord knows who is paying for it.

'Well,' she says, 'what do you think?'

I try to get my mouth to work.

'Tarte tatin,' I say eventually.

She looks away. Leans to pull some weeds from the base of the tree. Then she helps me back to the chair.

I keep a close eye on where Dano heaps the cuttings, and Saoirse tosses some of the rounds into his dimpled aluminium trailer – Bong! Bong! – even though I think she should not bend like that.

**II**

# Chinese Sunglasses

B y the time Monica found the long, narrow alleyway that was the Antique Market in Tianjin, it was well past lunchtime, and she was pooped. She would have loved to sit in one of those little restaurants along the way, but when she had tried that in Beijing, she had ended up with waitresses or waiters gathered around, dumbfounded, trying to be helpful. All she ever wanted was rice or noodles, and these were the only two things they never had a picture of. Street food was easier. Her favourite was roasted sweet potato, served whole in its scorched skin. You couldn't go wrong with that. But the day before yesterday she had eaten an iffy cabbage bun (only 2 yuan, or about 25 cents at home) which had left her queasy and without a solid motion ever since. Today, in the pedestrianized side streets of modern Tianjin, she had passed stalls that sold fabulous-looking street-food, but it was mostly skewered meat or squid, or very sweet pastry, and her stomach just wasn't up to any of that today. It was easier to just keep walking.

This was Monica's first real holiday abroad. You couldn't count the class romp to Malaga, after their final exams. Or going to Tenerife with her parents and brother. This time she was paying for herself, but she wasn't self-assured enough for all this sightseeing on her own. It had worn her down. Her friend, Sandra, had been the driving-force behind this holiday, but since they arrived last Thursday, until Wednesday this week, Sandra had to spend her days at a convention centre in the northern suburbs of Beijing, leaving Monica to her own devices. After Wednesday they would be able to explore together, and Monica yearned for that. A realization had grown in her: she needed someone to share her experiences. She had not known this about herself until she travelled so far from home. Nevertheless she was proud, in a silly sort of way, that she had seized this opportunity, and she was still amazed that she was actually here in China.

It was utterly different from anything she knew, in language, alphabet, food, culture, and yet they wore the exact same clothes and shoe brands, drove the same cars and used the same phones, probably because all those things were made here. The only negative was the smog – on windless days it was dreadful – and some of the public toilets were awful, but walking around as a single girl seemed completely safe security-wise. She had become gradually more adventurous, visiting the Forbidden City and the Temple of Heaven on her own, and even the 798 Art district, which had been really hard to get to. Now here she was in Tianjin, fulfilling the double goal of using the high-speed train and visiting a proper antiques market.

Monica rested for a moment. She hadn't understood before she came here that it would be so hot, and that clear, calm days would be the worst for smog, but it had been like this every day. Her throat was raw from breathing smog, and now, in these tightening old streets, there was the sudden additional smell of warm sewerage, like a pillow pressed on her face. She hurried along, trying not to breathe. Her stomach churned. She thought about the emergency tissues in her knapsack, hoping fervently that she would not have to use them. For just a moment, she felt truly weary.

On that night-out back in February with college friends, when Sandra had announced that Seedsavers was part-funding her to go to China as part of her organic seeds activism, Monica had said: 'Oh that sounds fantastic. I'd love to go to China.' And Sandra, with that gritty smile, said: 'Would you be free in June?' Monica hadn't expected that. She'd looked around the table at five wide-eyed girls, and Sandra – petite, ash-blonde Sandra, who would walk on burning coals to help anyone – looking at her with an almost pleading expression. Monica had only been working seven months at that time, as graduate entry to the audit section of Ernst and Young. She was barely entitled yet to any annual holidays, and although she had money of her own for the first time in her life, she still had to repay her college loan and kit-out the unfurnished duplex she had leased with Liz Tormey and Sheena Lynch. But she had the completely unjustified feeling that anything was possible. 'Yes,' Monica had said. 'I might be free.' There had been shrieks and gasps. No one was more surprised than Monica herself. Sandra had come around the table and hugged her, mumbled in her ear: 'You've no idea what this means to me.' Her embrace was like a binding of barbed wire, and Monica realised that Sandra had been in genuine trepidation.

It was a pity Sandra was a lesbian. Not that Sandra or any of the girls allowed that to affect their friendships, even though Sandra's partner Claire was about as friendly as a goat. But certain aspects of a normal holiday were obviously not going to happen here. Last week, after they had settled in their hotel near the convention centre, and made their way into the centre of Beijing for a quick recce, Sandra and Monica had gone into a western-style coffee shop, and two Australian guys had come in asking for directions to the subway. They were nice guys, tall and tanned and good humoured. They'd had a conversation, and possibly the guys might have asked to meet up some evening. Then Sandra said: 'It's particularly hard for gay people here. I find it repressive.' And she nodded emphatically. There had been a slight pause. The two guys had glanced at Monica, with her curly reddish hair and freckles, and she felt herself perform what her brother Mikey called the two kilowatt blush.

Since college, there had been a subtle change in the matter of men. There just weren't as many around. Surely it was not possible, simply not possible, that at twenty-four years old her best days were behind her. Eighteen year-olds looked now like willowy slips, and yet Monica could distinctly remember when eighteen had seemed ancient. She herself had definitely put on weight over the last few months – where had that belly come from? Drinks a couple of nights a week? Take-out food? It was so unfair – now that she could finally afford to eat and drink, she had to try not to.

The items that filled the stalls and little shops of the Antique Market seemed genuinely old. Everything had the patina of touch and use. Monica was fascinated by the tiny old books. If only she could read Mandarin. There were diaries, and personal breast-books, and The Sayings of Mao – the only English words she sawall day – from the old communist era. There were peculiar musical instruments, carved tobacco pipes, long metal opium pipes, spectacles, coins, all manner of quaint personal items. She was drawn to a display of worn wooden moulds for making Chinese moon cakes – a surge of pleasure at being able to recognise them. But the stallholder noticed her flicker of interest, and, like a spider, he began to emerge from the gloom within. Monica moved casually to the other side of the lane. The porcelain at the back of these little shops looked interesting too, but she didn't have any knowledge of such things, or not enough to pay serious money and then try to nurse some delicate vase back to her chaotic little duplex in Malahide. Plenty of trinkets and jewellery, strapless old Russian watches (what good were they to anyone?), ancient cameras, weird occult-looking compasses for fengshui; all these character-defining nuggets or prized possessions winnowed from a generation passed on. Now they would represent that bygone age in the new apartments of China, or be souvenirs of exotic holidays in the homes of travellers like herself. Monica moved along with that walk you had to have, the slowest of all walks without actually stopping.

Eventually she could go no further. It was too hot and she was too hungry. Thank heavens for the floppy sunhat, but she had no sunglasses; the case for her sunglasses had disintegrated last summer, and the lenses had been scratched. She had intended to get a cheap replacement pair in China, but had decided to wait until the end, see how much money was left. A torturous rumble made her stomach knot-up. She upended her water bottle into her mouth, savoured the luxury its cooling draught, and turned to retrace her steps. She had to find something to eat.

Really, she ought to be happy about being hungry. At the dressing table mirror this morning, her skin hung like pastry. Her Irish butter-fat was melting in the heat. Her hips felt softer, less turgid. It was great; she wasn't even *trying* to lose weight. Maybe if she stayed here for six months, she might finally approach that elusive shape that was somewhere between sexy and fit and happy.

A stallholder waved his hand, drawing attention to a pair of child's bootees. Monica groaned inwardly – surely he doesn't think I'm pregnant? She made that twisty shake of the head that means thanks but no thanks, and barely glanced at what he was holding. But... they were strange little bootees. She paused, to squint. Flat thin soles crumpled with age, made of felt or suede or something like that, the silken uppers reaching past heel-height, like two hands with the fingertips joined, and a very oriental-looking curl-up toe. But the opening seemed disproportionately large. Despite being finely made, they were awkward-looking for a child's shoe. Monica raised her eyebrows at the man, made a motion of rocking a baby.

He shook his head, then gestured with both hands at Monica, nodding vigorously.

'No no,' she said. 'Depends how much.'

But obviously she had misunderstood, because at that magic phrase, his face brightened.

'Toe hundred,' he said

Obviously he thought she was bargaining. She held up her hand as a stop sign, and started walking. He was beside her immediately.

'Waa hundred!'

She waved her hand. Stared ahead, kept walking.

'Fief tee.'

Fifty. Fifty yuan. That was… just over six euros. Still, a dinner for two cost less than seventy, when she and Sandra had managed to eat together.

'Fief tee, fief tee.'

She could afford that. It was something small and interesting. The souvenir-buying out of the way.

'Ok. Ok. Fifty.'

He bobbed and smiled and led her back to the stall. She paid, and started stuffing the bootees into her knapsack, with a vague sense of triumph at having something to show for the day, but slightly embarrassed at being so easily led. The stallholder began showing her postcards. She shook her head, but he insisted, tapping a postcard urgently, thrusting it in front of her. It was an old-fashioned sepia shot of a sour-faced Chinese woman, reclining heavily clothed on a day bed. He was tapping at the woman's legs, and Monica saw that the woman had only distorted stumps of feet, in shoes almost identical to the ones Monica had just bought. She stared. Took a bootee from her bag, put her fist into the opening. It fitted, with no room to spare.

'Oh my God.' Monica breathed.

She gestured at the postcard. 'How much?'

'Fief teea.'

Monica mutely handed over the fifty she had received as change from her hundred. To her surprise, the man gave her back thirty-five yuan.

'Oh…' she sighed. So much for thinking herself money-wise.

She wandered off with her change and her purchases. The slippers seemed to hum in her knapsack like a dark presence. What would Sandra think of them? Any hint of oppression brought Sandra close to tears. Monica walked faster. No, she had done the right thing, she was sure of it. Definitely. Empathy, that's what it was. Empathy was surging in her veins. She'd had no choice but to buy those slippers, though she couldn't say why right now. There was no way she could have left them, even if she'd known what they were. She began making a little story to make sense of all this, then abandoned it. Really, she was just so easily led. She imagined the woman in the postcard turning to look at her, scowling.

Her stomach rumbled again. And so hot. Each day, because of the smog, the sun did not cast a distinct shadow until nearly midday, but now she was walking directly into the milky afternoon sun. She could feel her jeans and T-shirt tacky with sweat. A crinkle of hair fell across her face; she wiped it back under the sunhat. Every unoccupied eye was following her, much more so than in Beijing, where at least there were other foreigners. Now she came to an intersection of alleys, and – thank

God for small mercies – there was some shade in the side-alleys. She had passed this intersection earlier and the side-alleys had not looked particularly interesting, but now she thought she might as well try them if only for the shade. Maybe there might be a tiny café with a waiter that spoke English, with a tuna salad or a wholemeal sandwich or… a bran muffin, a bran muffin with a cup of Barry's tea and real milk in it. Oh, dream on. Even the antique stalls were tapering out already. Instead, people were making things on the street; bracelets and bright jewellery and polished stones. A few stalls were selling modern junk. Maybe on the next street there might be a café. A shop would do. Anything at this stage. She glugged water from her bottle, which was getting low.

A pair of sunglasses caught her attention, the only modern glasses she had seen in this market. They were in an open box, nested in a small pile of household articles on a motorized cart. There was something ridiculous about these glasses – what was it? Monica leaned to look. Fashion sunglasses, awkward-looking and kind-of fancy, in a boxy aluminium case lined with purple velvet. She laughed. No way you'd break that case! It looked homemade by a welder. A tired-looking woman jumped to her feet to see what it was that Monica was looking at; she thought it was the hairdryer, and lifted that out. Imagine – in this heat! Monica suppressed a giggle. She waved – no. But she picked up the clunky sunglasses. Such a terrible knock-off, it was almost funny. Really homemade looking. At least they didn't have the gall to slap a brand name on it. The side frames were ridiculously thick, they were actually heavy, and – what on earth? – the bits that went behind your ears had dangly things attached, like the earphones you get with a phone.

'Fuiftuy!' the woman said breathlessly. 'Fuiftuy!

'No-oo thank you.' Monica put the glasses down and turned to move along.

The woman shook her hands in the air. She ran around the stall, pattered alongside Monica, bowing and holding out the ridiculous aluminium box with the glasses perched inside. She gestured all around, for Monica to try them out. Monica sighed. She did not stop walking, but to placate the woman she lifted the clunky glasses, and tried them on casually as she went. Good lenses, fair enough, dark honey-coloured. The earphone-things nestled at her ears, and there was a tiny click, a noise like an electronic chime.

'*Ni hao, ni jiao shenme migzi?*'

A female voice in the earphone things.

Monica stopped dead.

'*Ni jiao shenme migzi? Ni hao?*'

'Hello?' Monica said

'*Ohh!*' There was a scrambling noise. '*Ahh... Eengleesh?*'

'You're able to hear me?' Monica said. 'Who is this?'

From the little dangly earphones, there came a sound of confused Chinese arguing. Monica took off the glasses. Scrutinized them. Those little dangly things *were* earphones. She looked at the woman. The woman nodded eagerly: 'Fuiftuy!' and made a fist with one hand, five fingers with the other.

Monica poked at the earphone parts. 'What's all this stuff? These bits here. What's this about?'

The woman stared at the earphones in confusion, her head inclined sideways. She looked at Monica. 'Fawty?' she exclaimed hopefully.

'No no, what *is* it? And who was that... I mean...'

The woman continued nodding, nonplussed, offering the outsize aluminium case again and again. Monica held up a hand.

'Look, this! What's going on with this?'

'Tawty!' the woman said in desperation.

Monica shook her head. She put the glasses on again, fitted the earphones in properly. The electronic chime sounded again, right inside her head, it seemed, now that the earphones were fitted properly. She waited, hands to her ears.

'*Amm, excuse me, amm, Lady?*'

A man's voice. Monica froze.

'*Amm, hallo? Can you hear me?*'

'Yes.'

'*Aah, yo American?*'

'No. Irish. From Ireland.'

'*Ohh...? Eye land...? Aah, Ai'erlan! Aah! But I mean, amm, you speak English?*'

'Yes.'

'*Ohh. Thank you. Actually, I want know, how do you get these glass?*'

'I'm sorry?'

'*Amm, you say sorry?*'

'No, I meant I can't hear you. You're speaking too low.'

'*Oooh. Regret. I will... I will try more harder.*'

'Ok, that's better. I hear you perfectly now. What's going on with these glasses?'

'*Oooh! They are the glass of my company, they are taken of my boss apartment last night.*'

'Eh? I don't understand.'

*'Aam, my boss is… aww… test them. And they are taken of his apawtment.'*

'Listen, just a minute. First of all, who are you? And what're these … glasses?'

*'Amm, actually, my name is Wu Qiang.'*

'Woe… Chung.'

*'Aah… no, no. Wu. Qiang.'*

'Woo! Chang!'

*'Aah… well. Is too difficult for you. And my company name is… ah, you know this is the typical. I am only speaker of English in my company, and before I come, my boss has title for English as "Next Emancipation Electric", but in fact actually, the meaning of Chinese name is New Freedom Electronics, so I think I will tell you it is New Freedom Electronics, it is better.'*

'I see. Thank you, Woo, but–'

*'I mean, actually, my personal name is Qiang.'*

'Oh, ok. Chang!'

*'Ahh… no no. Maybe, actually, Chinese adopt English name. I did not speak with native speakaw before, so I have no English name, but if you want you can acall me Petaw.'*

'Peter.'

*'Exactly. So, now we are talking.'*

'Well, you're pretty good at English, Peter. Where did you learn?'

*'Ah noo. Not good. I teach myself, after school all myself. Twelve year, every day one hour.'*

'I see. You speak well.'

*'Oooh. No. No. I have no chance speak. You are first real person I speak.'*

'Well, it was nice to speak to you, Peter.'

*'Ahh, so kind. But aah, Lady? It is very nice speak you, but important matter is glass.'*

'Right… but they're nothing to do with me.'

*'But Lady, please, you seem nice person, and you have our glass.'*

'Yes, but they're not mine. I didn't buy them.'

The stallholder woman was staring up at Monica with open mouth, the aluminium case in her hand.

*'Um, pardon, lady? What you say?'*

'I said I didn't buy them.'

*'Oh Lady, where you are please? Can you tell me?'*

'Tawty!' said the stallholder frantically. 'Tawty!'

*'Uum, Lady! Who is speaking now? Who says this?'*

'What?'

'Tawty!'

The stallholder was shaking her fist in sequences of three. Now she was crossing her fingers, hitting them three times with the aluminium case.

'Lady…'

'Listen, really, I have to go. This is nothing to do with me.'

'Can you reach to glass, your right hand, small button is on top? Can you do?'

Monica had been about to take the glasses off. She felt along the overly thick side frame, and indeed there was a small bump. She pressed it. There was a SHICK noise. Nothing else happened. Monica waited, she and the stallholder agape at one another.

'One minute, Lady. One minute… Oooh! We have picture! You are in market! Oooh, terrible! What market you are—'

'Taw wenty!'

'Oh Lady! Lady, please! Do not give glass to that woman. She is robber.'

'But it's hers. She's selling it.'

The stallholder woman inclined her head one way, then the other, like a chicken, looking at Monica. Inside Monica's head, an excited conversation in Chinese was going on in the background, then the man's voice came back.

'Um… Lady. If you give her money to buy, we will give you back.'

'You'll what? Give me money back?'

'Exactly. And you give glass.'

'Oh, this is very peculiar…'

'But Lady, is very big service to me. Very appreciate. And not cost you any money, only do good service.'

Monica let her head hang. How did she get herself into these situations? She was so gullible. Sandra's guidebook had warned of elaborate scams to get money from unsuspecting tourists, but nothing as elaborate as this. For twenty yuan? Monica felt torn with indecision. But also she felt a tiny frisson of excitement. She thought about how much money she had in her purse. She had loads; about half of her remaining money, almost three hundred yuan, just in case. The daily budget was forty for lunch (€5) which meant roughly two hundred and sixty (roughly €32) for spending if she saw something nice. Of that she had spent sixty-five yuan so far. The subway in Beijing on the way back would cost only two yuan, to go the whole way around to their hotel in Hepinglu, and the high-speed train ticket was already paid for. So this was no big deal. She had missed lunch – she could afford to risk 20 yuan. It would be an adventure, as long as it didn't get out of hand.

'Ok. But I'm not giving her one cent over twenty yuan. Right? I don't have money to spare.'

*'That is wise, Lady. She is robber! But Lady, do not take off glass. We must speak, so we can make a… a… umm…'*

'Rendezvous?'

*'Ahh, excuse?'*

'An arrangement.'

*'Awwangement! Exactly! We must make awwangement.'*

The stallholder gaped at Monica with an expression of woe, but this changed to delight when Monica began to fish out her purse. Monica selected a twenty yuan note, and brandished it with what she hoped was an air of authority. No messing please. The woman nodded excitedly, took the note, and extended her hand for the glasses. Monica shook her head and pointed at the aluminium case. The woman handed it over. It was even heavier than it looked. There was an exact shape where the glasses would fit lens-first into the velvet, with a little metal connection in the centre. It was the geekiest thing ever. Now the woman was giving Monica a short black wire with a two-pin plug on the end.

'Eh, Peter? Peter?'

*'Oh yes, excuse. I forgot name.'*

'Is there supposed to be an electric wire attachment with the case?'

*'Yes. Exactly.'*

'What's that about?'

*'Aw, you know, that is my invention. There is charger in case. For transport chawge. Only small battery in frame of glass, for run phone.'*

'I see. Very techy altogether. Is this some advanced new stuff?'

Peter laughed, a little wryly Monica thought.

*'Aww, no, not advance technology. I make from broken cell phone. Just camera and some pieces, and aerial. Most expensive part is small battery. But battery is borrowed by my boss, so he must get back. Glass is my project, but he pay.'*

'Are you some sort of master inventor or something?'

*'Aw, I am engineer, but this is project by myself, hopefully for develop.'*

'Oh, ok. Well look, I've bought these glasses now, so do you want to come and meet me?'

*'Yes. Exactly. Boss will come and will give you money. He say he buy you new glass also.'*

'Oh. I see.'

*'Umm, Lady, can you tell me where you are present, boss go immediately find you. He is afraid for battery. But Lady! Important! If glass go quiet, must*

*put in box twenty minute, plug in. Understand? Otherwise, aww, do you
have phone?'*

Monica got that cold feeling again, of things moving out of control.
'Aah... I don't think so, not Chinese phone, sorry.'

*'Ok. Twenty minute for glass in box, then we can talk again.'*

'Ah... well, I think I know what you mean. Charge it up, right?'

*'Exactly. You are clever.'*

'No I'm not. I can't even get something to eat.'

*'Why so? You get for eat and wait boss.'*

'No no. I just can't order.'

*'Can't odour...?'*

'Cannot speak. Cannot say what I want.'

*'Ohh. You go restaurant with picture food. You pointing.'*

Monica laughed at the craziness of it. She felt like a child.

'But they always bring what I don't want. Or something's wrong.'

*'Ohh...'*

Monica took her sunhat off, and the sunglasses. She shook her hair,
ran her fingers through a couple of times. She put the sunhat back on,
then the glasses. The electronic chime sounded.

*'Ahh, Lady. You went away.'*

'Just had to settle myself a little.'

*'But you don't say which market are you? This moment.'*

'I'm in the Antique Market in Tianjin.'

*'Tianjin! Ohh!'*

More animated Chinese conversation in Monica's head. There seemed
suddenly a sense of danger. Like she was involved in something illicit. But
she was blameless: what had she done wrong? She turned, and despite
the heat, paced back to the main thoroughfare of the market, turned in
the direction of the modern shopping streets. Again the curious stares
of stallholders, the ones not playing games or watching videos on their
phones. One pair were playing something like draughts, with a small
crowd gathered around. It was good to be among people.

*'Umm, Lady!'*

'Yes?'

*'It will take boss one hour and half for get Tianjin. Please, is ok?'*

'Where are you? Your company.'

*'Ahh. Yes. Actually, you can write?'*

'Yes... oh I see, good idea.' For some reason she had expected that he
might not want to give an address. 'Hold on... till I get a... pen.' She
dug her diary and a pen from the side pocket of the knapsack. 'Ok.'

*'You remember, for address must be called Next Emancipation Electric, and situated... umm... I think in English is call Muning Load – M-U-N-I-N-G, Number 21879, Teda, Binhai. So. You know where glass belong.'*

'Thank you. If anything goes wrong with your boss, I'll post it. Ok?'

*'Ohh. Thank you. So kind. I think you are honest person.'*

'Yeah, that's me.'

*'Now, we must arrangement.'*

A giddy feeling came over Monica. Perhaps light-headedness from hunger. She felt an ability to be daring, that she had left her personal shore and could strike off now in any direction.

'Peter, if I go to a restaurant, do you think you could tell me what to ask for?'

*'Tell you... aah... what...?'*

'Translate. Translate what I want.'

*'Ohh. Of course. But sorry, boss will go Tianjin. Where will you meet?'*

'Oh, I don't know. Any place. We'll choose a restaurant or coffee place maybe, when I get in there, ok? And I'll wait for him.'

*Ohh. Exactly. So you will go to centre now?'*

'Yes. But first, something to eat.'

Monica began walking with a purpose.

'Peter.'

*'Yes Lady.'*

'My name is Monica.'

*'Moon uca. Is right?'*

'Yes. *Mon*-i-ca.'

*'Moon y ca. How spell?'*

'M-O-N-I-C-A.'

*'Ahh... Moan-y-ca.'*

'Yes. Now, Peter, tell me, are you able to see what I'm looking at?'

*'Ah, noo. Must press button to camera, then picture come to me on computer. That is important function of glass.'*

'What do you actually do, your company?'

*'We are making apes.'*

'Apes?'

*'Yes. Apes for cell phone, and small device.'*

Monica smiled, and nodded to herself, thinking of the time she had repeatedly said 'Noodles' to a gob-smacked waitress, until she heard the kitchen staff shouting New Dolls! at one another and cracking themselves laughing.

*'Soon I hope we are making inter-active glass, so person in our office can*

*give help. Can use internet for person using glass. Is good service, yes?'*

'You could do travel advice. Translation.'

*'Exactly. This what we develop. We do all things for customer.'*

'Well, you're going to have your first customer in one minute. Can I take a picture?'

*'Of course. Certainly!'*

Monica had stopped in front of a small street-side restaurant, much like any of the others, that had large paper lanterns and flashing neon lights and bright red Chinese lettering which she could understand nothing of. She pressed the button on the right hand frame. *SHICK.*

'Ok, will you tell me when you get that?'

*'Ok.'* Some seconds passed. *'Yes, I have. Is restaurant. Local.'*

'Right, well I want some sort of cooked snack, ok? I don't want any meat or prawns or seafood or especially not cabbage, ok? And no salad. Just something simple, not sweet, and sit down. What would they do like that, d'you think?'

*'Oh they will do lot of things. Go look at menu. You know? Steady with head and glass, take picture. Make sure glass do not move on head.'*

Monica went up two steps and pushed through a screen of plastic flaps hanging in the doorway. She found herself surveying a square space with a yellowish terrazzo floor, modern tables of thick Formica, and art deco style chairs of black lacquer, all a tad shabby.

'Oh oh, here comes the waitress.'

*'Take photo.'*

One entire wall was covered with lurid photographs of food, each with something written underneath in Chinese. Monica nodded to the waitress, pointed at the wall. She put both hands to the glasses, and took a steady stance. *SHICK.* She felt full of confidence. It was a great feeling. For the first time since she arrived, she felt in some sort of control. She had a friend. A tiny shiver passed through her.

*'Ahh, Moan-y-ca. I will say now. They have a steam a dumpling, a steam a bun, a noodle with soup, a soupa dumpling, a pancake with—'*

'Pancake! I'll have pancake. How'll I get that?'

*'Aww, you can point... let me see. On wall, you see top, the fourth column, second from top?'*

'Fourth column, second from top?'

*'Yes.'*

'That's prawns.'

*'Naw! Let me see... Ah, prawns is other side. Fourth column I say, not fourth from end.'*

'Oh I see. Fourth from right you mean.'

*'Yes. Or you can say "Jian bing quozi".*

'Ten bean quartz ill.'

*'Well... one time again. But try... aah... get tone. Jian – bing – quozi.'*

'Ten bean *quartz* ull.'

*'Oh, ok. You are speaking Chinese.'*

Monica turned to the open-mouthed waitress.

'Ten bean *quartz* ull.'

The eyebrows did a little jump, and a stream of Chinese came out of the girl's mouth. She flapped a hand at the wall. Monica felt prickles of heat on her scalp. She waited rigidly for Peter to say something, and when his voice came, it was like a waft of cool air.

*'Ahh, she is complaining small bit, wants know if like anything except than egg inside it."*

'Egg?'

The waitress's eyebrows went up when Monica spoke.

*'Yes, pancake is cooking with fresh egg. Very good. But can have other thing too.'*

'No, nothing else.'

The waitress was jabbering worriedly, looking toward the kitchen for assistance.

*'Ok. Say "Bu".'*

'Boo!'

The waitress jabbered again.

*'Anything for drink?'*

'No.'

*'Say "Bu".'*

'Boo!'

*'Want sit in or walk away?'*

'Oh, sit. Sit.'

She sat at a table halfway across the room from the door, the only customer.

*'Moan-y-ca, you should take off glass, save battery. While you eat.'*

'Ok, see you later.'

*'Ahh, what you say? See me?'*

'I mean goodbye for a while, talk to you later.'

*'Oh yes. We talk soon.'*

Monica let her arms flop. The stickiness of her clothes was ridiculous. And it must be nearly seven hours since she had eaten. A gorgeous smell of frying wafted across the room, the aroma of fresh coriander and parsley and spring onion. Saliva gushed in her mouth. Her stomach rumbled loudly.

The pancake arrived, a large crepe really, folded thickly into quarter size. Monica devoured it with her eyes before the plate touched the table. She lifted it. It was like flat bread, a little oily, with an egg cooked into it, just enough coriander and onion, and some kind of hot aromatic spicy sauce. It was gone so quickly that she called the waitress, and by holding up a finger and pointing and nodding, managed to convey that she wanted another one. This successful communication gave her a warm feeling. The second pancake was brought to her, and she ate that. A wave of contentment seeped down through her, out into her limbs. She sank back in her chair, sighed, wiped her mouth with a tiny tissue from a box on the table.

Now the bill. All those evenings trying to learn phrases had been a complete waste of time. Even when she used the correct phrase, it only served to create hilarity if they discovered afterwards she had been trying to speak Mandarin and had been saying the right thing all along. Now here came the girl bringing a piece of flimsy pink paper with some illegible scrawl on it, and no pen.

'Shi-liu!' said the waitress, making a sort of hand signal like news for the deaf. Normally, Monica guessed a denomination of banknote that would be safely above the price, but she suspected that this might be encouraging inflation, especially in the fruit markets. 'Shi-liu!' said the waitress again, tapping at the paper.

'Hold on.'

Monica put on the sunglasses.

*'Ah hello Moan-y-ca. Are you finish?'*

'Shi-liu!'

'Yes, thanks. What is she saying?'

The waitress's jaw became slack. She cast another pleading look toward the kitchen.

*'She is saying sixteen yuan. Did you have two pancake?'*

'Yes,' Monica said guiltily. 'They were great.'

She fished out notes, and included a tip of a one yean note. The girl gave it back.

*'Moan-y-ca, you go in shopping street now?'*

'Yes, I go.' She shook her head at how she was beginning to speak in stilted English herself. 'Is it best if I go and find a place and then talk to you?'

*'Yes, boss will be an hour more. He coming quick, but long train.'*

The street outside seemed brighter, more optimistic. Monica's step was light. She slung her knapsack on one shoulder and paced along, noticing little houses behind the shops and restaurants, some with traditional curved-tile cappings on their yard walls. Then she came abruptly to the first of the colossal new buildings of the modern shopping street, with their sharp modern lines and huge concrete swellings and Givenchy adverts. The sophisticated 21$^{st}$ century effect was tempered by street vendors spreading clothes and other wares on the new walkways underneath. The European models in the Givenchy advertisements were hollow-eyed and stick-like, pouting bad-temperedly at the street below. Probably the Chinese had thought everyone in the west looked like that until she'd turned up.

'Get a life,' she said to the adverts. 'Get a sandwich.'

She still felt slightly guilty about eating two pancakes. When she and Sandra arrived in China last week, their first impression had been that every single person was fashionable and fit looking. Certainly everyone on the Beijing underground. Even the older women were like chic Parisians in natty little dresses and suits. Sandra fitted right in, except for her ice-blond hair and her penchant for long skirts and Doc Martins. Monica had felt like some kind of two-legged mastodon, escaped from the wilds of north-western Europe, and needing three meals a day. Within an hour she had mentally abandoned all hope of finding a silk jacket that would fit her. Since then she had probably seen about a million people, and only two of them were actually fat; she had noted them specially. What the hell, maybe they couldn't afford food with all the money they spent on phones. She jaunted along, toward the main pedestrianized shopping street, noting particularly all the fashion items that only slim people could ever wear.

She came to a long sweep of futuristic building design that could have been an airport. Where would she find a coffee shop in a place like this? She looked at her watch. Less than an hour before boss-man arrived. It was all slightly thrilling. Like an espionage movie. Wait until she told Sandra. Although Sandra was hard to interest in ordinary affairs. Maybe this seemed like an adventure only because she was alone. Hopefully nothing would go wrong. She felt light, relaxed. Things were going well today, after the gloom of the stomach upset.

Monica slowed down because she was getting overly warm again. The crowds had become dense as she moved along the wide pedestrianized street, and every single person seemed to notice her. Not that they exactly stared, but their initial expression of surprise would slacken, they would drop or avert their eyes, become stony-faced until she passed. Anytime she cared to, she could look behind and see someone sneaking a reverse peek. Heavens knows what they thought of her. Probably they were thinking – where does she get clothes to fit her? Good question. Nowhere around here, that's for sure. Although maybe a bit of insider information would help. Aha –now there was an idea.

She took out the glasses.

'Hello, Peter?'

It was amazing how she looked forward to his answer. As if the whole street, the sky even, was different because a Chinese man called Peter, whom she had never met and knew nothing about, would answer this... this call.

*'Ah hello Moan-y-ca. How are you? Are in city?'*

'Yes. I will look for a place to meet your boss.'

*'Aah. Certainly. So kind. But aah, Moan-y-ca, before boss come, I want ask, what sort person are you?'*

She felt that tiny shiver again.

'Ah... jeepers, I'm not sure what you mean, Peter.'

*'Well, I mean such lucky you find glass. Is like...ooh... fairy tale. I thought you must be rich and beautiful.'*

'Haa! Afraid you got the wrong gal this time, sorry. Just ordinary... plain person.'

*'But you are so kind and... umm... decent. You do not try to steal. Instead you help us.'*

'I don't mind helping. Long as it doesn't cost money.'

*'But how you come China? Is much money take plane.'*

'What about you, Peter? Are you rich and beautiful?'

It was like she was suddenly brave and daring, or that Peter didn't really exist or it didn't matter what she said to him. And it didn't, really. But now he was laughing, almost falsetto. He seemed to have to explain to someone why he was laughing, and now other people could be heard laughing too.

*'Ohh, sorry. So sorry. Aam... aah... no, no Moan-y-ca, I not rich.'* He controlled himself, his voice became lower. *'Actually, even boss not rich, I think.'*

'Never mind that, are you good looking?'

Monica could hardly believe she said that.

*'Oh?'* He laughed again, that sudden high pitch. *'Ahaa! Noo! No, not good looking actually.'* He seemed bashful now. *'Aah..., I not good to look. I... ah... too brown colour.'*

'Too... brown? But you're Chinese, right?'

'*Yes, yes, Chinese. I from Gansu Province. You know?*'

'Eh, no, sorry.'

'*Is western Province. Near Qinghai and Mongolia. But here they say too brown. They... ahh... call me Muslim. Or farmer.*'

'Are you Muslim?'

'*No no. My family Buddhist. But my friend Muslim. I am from village near town of Xiahe, has very big Buddhist... amm... amm... place. Big... aam... temple.*'

'And you come here to work?'

'*Yes. First university, then work.*'

'D'you mind if I ask, what age are you?'

'*Me? Aww... twenty eight... aw... years.*'

'Oh, you sound, well, younger.'

'*But Moan-y-ca, you... aww... sound young also. How many years you?*'

'Now Peter...' She was about to say you don't ask a girl that question, especially after a certain age, like twenty-two. 'I'm twenty four.'

'*Oh Moan-y-ca, you so wise, and so young. I can't believe. Is it true?*'

'I'm twenty-four, that's true.'

'*Aah. And excuse ...umm... are you married?*'

Monica laughed. 'No, are you?'

'*Aww... no. I had girlfriend, two girlfriend, but is got married some one else. I have no apartment and no car, so cannot get wife. Maybe two year I get apartment, and three year, get car. Then can get married.*'

'Well, I've a car, and I'm not married. Marriage isn't everything, you know.'

'*Aww! Woman is always too choosy! Especially woman with car. What kind car you have?*'

'Renault. Renault Clio.'

'*Ooh! Strange car! Where is made?*'

'France.'

'*France... France? Oh, so expensive. And you apartment, how many metres?*'

'Metres? Haven't a clue. It's a duplex. You know, on two floors. Two bedrooms.'

'*Ooh...*'

He sounded awed. Probably imagining some gorgeous place overlooking the sea, not a hastily finished development with cheap pathways and a tiny green that cost a fortune to maintain, paper-thin walls so you can hear the neighbours on three sides arguing and watching TV and going to the toilet, sometimes all at once.

*'Moan-y-ca, would be possible, you find mirror, take picture?'*
'Take picture?'
*'If you don't aw mind.'*
'A picture of me? Why?'
*'Well... I wonder a lot what you look like.'*
'Peter, believe me, you're better off using your imagination, ok?'
*'Oh, apologise. Is ok.'*
'I'll tell you what, Peter, I'm going to shop for a silk jacket, and if I find one that I can fit into, I'll show you what it looks like on me, ok?'
*'Silk jackah...?'*
'You know, a Chinese jacket, high collar, with little clasps instead of buttons?'
*'Amm... not actually.'*
'A formal jacket. For wearing out. A traditional jacket.'
*'Aah. Maybe costume of Tung dynasty.'*
'Maybe. I don't know. I'd have to see.'
*'Wait please. I will find for you.'*
Monica could hear discussion amongst several voices. Peter was being forceful, but the street noise around Monica was overwhelming, especially the repetitive electronic megaphone-shouting at nearly every shop doorway. To cap it all, loud music burst abruptly, crazily, from huge speakers in an air-pumped arch of sky-blue colour. This was a marquee devoted to Aupres cosmetics. The windows of the department store behind were also full of Aupres products. How much perfume would they have to sell to justify all that? But now a lithe young woman in a pink sequined mini-dress had begun playing an electric violin with great gusto, at the same time dancing in gymnastic fashion. Mother of God! Probably she could cook and read a book while she was practicing that at home. The act built-up to a shrill, contorted crescendo.
*'Moan-y-ca! Where are you this present? Moan-y-ca?'*
'Sorry, can't hear you! I'm watching an eel in a mini-dress.'
*'Aaw...pahdon?'*
'Oh nothing, nothing. Just really loud music. Let me go over here where it's quiet.'
*'Aww... Where are you?'*
'I'm... Golly, I've no idea. I'm on a pedestrian street in Tianjin.'
*'Aww, good. So, take picture. Street picture.'*
Monica put her hands to the glasses. Found the button. *SHICK*. And waited.
A muted chorus of opinions behind Peter.

*'Aww, actually you are on Binjiang Dao. Is perfect actually. Now continue direction, four street, make turn right. Take photo for confirm.'*

'Why would I do that?'

*'For place make Tung jacket. Silk.'*

Monica's eyes widened. She hesitated a moment, not really thinking about anything. But then she began walking. She paced along with purpose, trying to keep her excitement in check. But was she just being led again? She was torn between curiosity, and wanting to trust Peter, and fear of being led astray. She stopped at the fourth turn, took a picture, and Peter gave directions that took her down a side street away from the large shops. It seemed to be an area of small businesses, built onto the fronts of houses. You couldn't tell what some of them were selling, but it definitely wasn't mainstream commerce. Suppose he led her someplace off the beaten track, set her up to be mugged? Anyone looking out for a foreigner would see her bushy mop arriving half a kilometre away. Monica put on the glasses.

'Peter? Hallo?'

*'Yes, Moan-y-ca?'*

'We seem to be going into residential places here.'

*'Yes. You need pass next street.'*

It was in an old part of town, like the Antique Market. Two story buildings almost to the street, a narrow higgledy-piggledy pavement with trees blocking the way and cars parked everywhere. You couldn't see what was going on. Some of the buildings were shops, and some seemed to be just houses, others were god-knows-what, she couldn't see beyond the narrow iron gates. She looked back, to memorize where she had come from, and moved tentatively on. She came to some small modern shopfronts of plate glass, with low walls separating them. At least you could see into these; the first was a beautician, with women inside being dolled-up; then a clothes shop for children; then a shop selling nothing but golden Buddhas for heavens sake, and… a clothes shop that sold silk dresses and a brocade jacket! Oh, it was closed. But in a glass display outside, there were three of the most shapely Chinese silk dresses you could imagine, and a brocade jacket. Exactly what she wanted. But no prices… except on one red and pink dress. 1990 yuan. Shit! That was… about €250. In fairness, only a film star could wear a dress like that. It was beautifully made, fabulous shapely cut, and those flowers around the neck must be hand embroidered. Wow. Still, it wasn't cheap. But why couldn't they put a price on the jacket too? Blast them.

She fumbled for the glasses.

'Hi Peter.'

*'Hi Moan-y-ca.'*

'Is this it by any chance?'

She pushed the button. *SHICK*. Then in her head she could hear a discussion, an argument.

*'Moan-y-ca, I must a send picture someone else. Please wait.'*

'Ok. But what does that sign say on the door? The red sign.'

*'Ooh, it says Closed.'*

'Great.'

She waited.

*'Oh, I have answer. One moment.'* Peter began speaking animatedly. Accepting some fact, by the sound of it. *'Moan-y-ca, is not right place. Is too expensive. You do not reach next street?'*

'Next... street. No.'

*'Next street, cross over. Take picture.'*

She walked about fifty metres to the next intersection, crossed over. She put on the glasses, and when there was no traffic she stepped out, and took a picture.

Waited. A discussion was taking place in her head.

*'Hao. Hao. Ah, ok, Moan-y-ca. Very good. You have made it. You see on the left maybe some metres, is red lanterns. Yes?'*

'Red lanterns. Yes, I see them.'

*'Ok. Before, two building, is entrance shop.'*

'Wait, wait. What? Before the restaurant, or after it.'

*'Before! Before! Two building.'*

'Ok. Hold on a minute.'

She walked right up the restaurant, just to be certain, counted two entrances back. The first sold umbrellas and purses and belts; accessories basically. The second seemed to be a private house. A metal gate was open, leading into a short narrow alley. Exactly what she had been afraid of. This was no shop. She looked at the next doorway before that: it sold tubes and cans of some sort of lotions, from shelves on the pavement. That wasn't it either. Perspiration grew warm beneath her clothes.

'Peter, there's no clothes shop here. It's a house or something. Are you sure this is the place? Do you actually know it?'

*'Aah...no, I don't know. I send picture to lady in next business beside here. She know area, she know good clothes. Says is right. You try?'*

It was dark-looking, spooky. She didn't want to say the words: I'm not going in there.

*'You go. Yes?'*

Monica felt her teeth compress until her jaw muscles knotted. Heat wafted up the neck of her t-shirt. She stepped over an ankle-high threshold stone, hands outstretched as if to fend off some unseen obstacle. Tried to be silent. A white and grey cat was sitting complacent and smug on a low roof. It's head rotated to follow Monica's progress. She crept warily along the flagstone path. It was a really old place, with clusters of dirty wires hanging everywhere. A vine and sort type of leafy climbing plant grew onto a metal trellis from large earthenware pots. And there were spikes on the walls. Thin metal spikes. There were even spikes on the ridge-tiles of the roofs. Not exactly spikes, but metal things. The windows were narrow and translucent with dust. Oh, why didn't she check the street before she came in? She should have checked the street. This narrow place was too quiet. Filled with clutter: wires, pipes, crates, bicycles. The city noises seemed muted, far off, and what if that gate clanged shut? She had tip-toed almost as far as a columned portico. The door was open to a dark hall. No way was she going in. All she could see in there was a wall, half hidden by cardboard boxes and cylinders of some sort. Peter's voice, suddenly, made the hairs on her body stand on end. She snatched off the glasses, stuffed them awkwardly in her jeans so that she could see better, and concentrate. Pushed her hair back with both hands, to let the heat away. Knocked tentatively on the open door.

Nothing.

She knocked louder. Nothing.

Looked back toward the gate; so far off, although just a few steps. If only Sandra was here, Sandra was brave. Or decisive at least. Monica plucked again at her t-shirt. Leave. Now. Nobody would know. It just didn't work out. She looked toward the street. An electric scooter whirred past on the tarmac outside. An old woman with a toddler passed on the footpath. Monica longed to get back to the street. She shouted a tentative 'Hello?' into the doorway, ready to sprint.

There was no response for a moment. But a murmur became audible, then a clattering, as if a door inside had been opened. That noise became abruptly muted, and a wooden door swung into view. A short stocky woman entered the dark hallway, her mouth making an "o" when she saw Monica.

If it had been a man, Monica would have muttered "Sorry!' and fled. She might have done that anyway, if the woman had not been wearing the most understated and elegant silk jacket. Monica paused in her ready-to-sprint attitude, and pointed hesitantly at the silk jacket. It was gorgeous. In the poor light you couldn't tell if the colour was green or grey or palest blue, and the pattern swam before Monica's eyes. But she could see clearly that the style of that garment made this stocky little woman look like an empress. It had beautiful brocade double buttons, a high collar, and long elegant sleeves. You couldn't focus on any aspect because it didn't have a flaw.

The woman was possibly in her late fifties, wearing just enough make-up to look sophisticated. She asked Monica something in Chinese, her eyes flickering with doubt. Monica pointed at the silk jacket, then at herself, nodding her head and raising her eyebrows as she did so. The woman's expression cleared. 'Aah!' Her face became business-like; she turned into the gloom, beckoning Monica to follow. Monica stepped cautiously into the hallway. She followed the woman through a wooden door, then a heavy sliding door, which was covered with silk dresses on hangers that swung to and fro.

Behind the door was a workshop. A sewing machine ceased clattering as Monica entered. Besides the stocky woman, there were four, six, seven women in there who became wide-eyed and still. Fluorescent strips high-up and at head level provided a glare of light onto two very large tables strewn with patterns, and two smaller tables with sewing machines. There was cloth everywhere, with a great emphasis on red, thin rolls of it standing in leaning piles, swatches hanging on poles and lines, flattened bolts heaped on cupboards and shelves. Pigeonhole storage covered nearly half of one wall and all of the space under the big tables. A pillar in the centre was festooned with scissors, measuring tapes, coils of braid and piping. Trays and plastic tubs of buttons or fixings or spools of thread cluttered every flat surface. It was like a cave. The women all had their hair in buns, except the first lady. Two were quite young, one very old, all in various types of tracksuit or work suits. Only the first woman wore a silk jacket, but Monica could see two more of these hanging up, one on an adjustable standing dummy. The women one by one began to congregate where the stocky lady led Monica. A tall lamp was brought over and some space created by moving high stools and a clotheshorse. The women talked all at once, softly, pointing at things and gently nudging one another.

In a calm flurry, they brought different sorts of jackets and dresses for Monica to see. Everything was beautiful, but Monica felt sick in her heart looking at the tiny size of them. She shook her head at the dresses, making a wry expression and patting her belly. The women didn't pay any attention to this at all. They argued together and softly sighed, parading different garments one after another, holding them against Monica, and taking them away or leaving them until the clothes horse behind was heaped. Eventually, Monica managed to convince them to send all the dresses away, there was no point in trying to fool anyone. The women protested and looked calmly disbelieving, but they took them away. Soon there were only jackets of various types and lengths. It became like a waltz, with slightly different styles and weights, linings, cross-overs, and buttons, all parading before her and swishing away. Some had different sleeves, others had different ways of ending the length, different waists, different style of slit at the side. Well-thumbed catalogues with bamboo-thin models were produced, pages flipped, details pointed at. The stocky woman in the silk jacket kept up a constant mutter, motioning women to get something else, or to display a different aspect of what they were holding. Finally Monica came to a conclusion about what she would like, and the cloth she would like, and the lining, while the stocky woman made notes and tiny sketches. And a terrible thought struck Monica. How much?

She became instantly too hot. Should have asked that before she started all this trouble. She held up a hand to indicate Stop. Tried to beckon the stocky woman aside, to save embarrassment, show her how much money she had. But they all came to look, so she gave up rooting in her knapsack, instead she rubbed finger and thumb together to signify money, raised her eyebrows, pointing at the sketches the woman had made. The woman said something with a little upward motion of her head. Monica raised her palms, and did a mime of writing something down. While the woman was doing that, Monica calculated furiously: she'd had three hundred yuan, give or take one or two; sixty five already spent, plus twenty, plus sixteen for pancakes, plus two yuan needed for the subway... she had about a hundred and ninety-seven. She looked at the woman in trepidation. The woman handed her a piece of crinkled paper. It had the number 250 written on it.

Embarrassment caused the blood to pound in Monica's ears. It was so little. Two hundred and fifty – she could no longer think – was that even fifty, even forty euros? It seemed like nothing to pay for a jacket like that. Yet she could not afford it. How did she get in this situation? It would be criminal to haggle about that price. What could she do? Have one without lining? Change the whole thing? Her face was probably crimson, sweat seemed about to trickle down her cheeks. Monica bowed contritely. Dug out her purse and opened it for the woman to see. She wrote on the paper: 197. The woman made a dismissive gesture. She put a line through 250, and wrote 230. Monica shook her head sadly. Felt her lungs vibrate. Get a grip, she told herself, for the love of heaven. She pointed to the 197, and made an apologetic gesture. The woman shook her head resolutely. She handed the paper and the pen to Monica, and invited her to write. Monica would cheerfully have written 300, but she could not do that. She looked at the woman, and had to clench her teeth not to become tearful. Perhaps pay just a deposit? And hope to get that measly twenty back from Peter's boss? Or get the train out here again, which would cost an extra hundred and ten. If she did that she would have to borrow from Sandra – who had even less money than she did, although she hadn't spent any yet. No. It had to be Peter and his boss. Peter said he would pay, and that better be true. Otherwise no glasses. If she could make it to 220 yuan, and the woman accepted that, she would get the train back to Beijing and walk across the city if necessary, if she did not have even two yuan for the subway.

Monica wrote 220 on the piece of paper. Looked hopefully, apologetically at the woman. The woman pursed her lips, and nodded. She put out her hand for the money. This was really going to happen. The woman was already giving instruction to two of the younger ones. But as Monica counted out money, she saw her train ticket. Oh no! She hadn't even thought about that. If they were going to make this thing from scratch, she'd have to come back and collect it after all. She took out her train ticket and showed it to the woman, pointing out the date and time. The woman took reading glasses from a pocket and considered the ticket carefully, then from a high shelf she fetched down a plastic old-fashioned alarm clock, showed Monica five o'clock, and pointed at the floor: ok, here, five o'clock. Then she turned her attention toward Monica's money. Monica counted out a hundred and ninety. She gave this to the woman, and while the woman was counting it, Monica wrote on the paper 190 + 30. She pointed at the clock and the thirty. The woman made a resigned shrug, clapped her hands and motioned the two women to begin.

They wanted her to take her t-shirt off, but Monica declined, several times, causing them to shake their heads and look at one another. They began measuring her, getting her to put her arms out and her chin up, and stand properly upright. They measured her waist and her hips, her shoulders, her bosom, over and under her bosom, her front and back, her arms extended and not extended. They murmured and sighed to one another, but when they were reviewing what they had written down, they broke into a giggle. What? She just had a little bit of a belly, that's all. Probably the first time in their lives they had seen such a thing. The stocky woman drifted serenely over, keeping an eye. When the measuring was over, the stocky woman pointed at the alarm clock, and without change of expression she slid the heavy door open and made a paddling motion with her hand, urging Monica to leave.

Monica entered back into the world of horns and motorbikes and brakes and revving engines. The milky blue sky glared through the trees and wires. She took a moment to catch her breath, to convince herself of the reality of it, and began striding back toward the pedestrianized area, dodging between trees and holes in the pavement and parked cars, her hands on the straps of her knapsack. It was good to walk, to settle things in her head, and feel the air moving around her. There was no thinking to do, just find a place to meet Peter's boss.

Only, she couldn't hang around in a coffee house for forty minutes without buying something, and she definitely could not buy anything – such a ridiculous situation to be in. What kind of eejit flies halfway around the world without a spare cent to splash out on a jacket and a cup of coffee?

She reached the pedestrian street, which even on this weekday afternoon was a river of humanity. The energy of them all, so young, a mass of expectation seething along. It would be nice to sit and watch, let them all flow past. Perhaps she could find a public seat. She drifted along, gazing into shop entrances of polished stone and hanging crystal and sharp white lights, and, to spoil the effect, raucous tinny loudspeakers at the door. Finally, some public benches in the centre of the thoroughfare. But they were filled with old people, the only older people she had seen, with sagged shopping bags in either hand, waiting for the energy to go on.

A sign for Costa Coffee. That would do. It was upstairs in a department store. She checked her phone. Ok, still some time before Boss Man got here. She put on the glasses.

'*Oh Moan—y-ca, I thought you lost.*'

'No, no. Busy, that's all.'

'*Do you find shop?*'

'Yes.' She took a deep breath. 'Yes I did, Peter. And thank you, you did a great job there. I really hope this business goes well for you.'

'*Ahh. So please! Is my idea.*'

'Your idea?'

'*The glass. For bring to work Gansu province.*'

'I'm sorry?'

'*Worker in Gansu answer glass, give support by internet. Is way of getting work in Gansu. Small money in Gansu can have good life. Be happy. So. Now you can wait boss?*'

'Yes. Lemme just take a picture of this place, you'll see the Costa Coffee sign, and I'll meet your boss there. Ok?'

*SHICK*

*'Yes. I see. Is department store. Binjiang Dao, yes?'*

'Where I was before, the same street.'

*'Very good. I think boss will be there. He phone say six station more, so maybe one half hour. He must get off train station, get taxi'*

'You mean half hour? Thirty minutes?'

*'Exactly. He will be one half hour.'*

'Ok. Ok, yeah, that's fine. And you told him the glasses cost me twenty yuan, right?'

*'Correct. He give you twenty yuan. So now. Have coffee. Wait.'*

Monica tensed.

'I can't have coffee, I've no money left.'

*'No money?'*

'No. But I have train ticket, I mean, I have *a* train ticket. To Beijing. But I can't buy coffee today.'

*'Boss will buy coffee. I phone he.'*

'No no, look – I don't want him to buy me coffee, thanks. Coffee's twenty or thirty yuan, but I do need just an extra two yuan, for the subway in Beijing. Do you think he could give me that? I know it's awful, asking for a bit extra.'

How crazy it was to be short of money. Especially such a tiny amount. Like last Christmas, she had given change to a charity on the street, and then didn't have quite enough for her fare home, just by a few lousy cents. A few cents short was the same as broke.

*'Certainly. He will be glad. But he also can buy coffee, is no problem.'*

'No no, you're too kind. I don't want to be any trouble. I'll just go for a walk, and I'll come here when it's time, ok?'

*'You will to shops?'*

'No Peter, the last thing I want is shops. Maybe there's someplace else around here for tourists.'

*'Oh yes, wait! Wait! I will find.'*

She waited there in the middle of the street, a rock in a river of people. Huge adverts on video screens blinked mercilessly. But she felt ok. She met the stares without fear, because she was waiting to hear from a man she had never met.

*'Moan-y-ca?'*

'Yes.'

*'For one half hour, you can go China House Museum. Take picture your*

*camera.'*
    'Great, what kind of museum is it?'
    *'China. China museum.'*
    'I know, but what aspect of China? What subject.'
    *'Aww... Chinese china. Old and broken maybe.'*
    'Aah!' Monica almost giggled. 'How do I get there?'
    *'Is two block, away from direction of market. Street is name Chifeng Dao.*
*You go Chifeng Dao, turn left. Maybe three, four block. You will see.'*
    'Thank you, Peter. That's great. I'll be back here at 4.30. Ok?
    *'Ok. See you later, Moan-y-ca! Ha ha – is right? Even I don't see you.'*
    Monica laughed. 'Yes, that's right.'

There was an entry about the China House in the guidebook, and Chifeng
Dao was exactly where Peter said it would be, which was for some reason
surprising and reassuring. Turning left, she dawdled blissfully for a few
blocks that seemed strangely familiar. They were western buildings. This
could be London, or even parts of Dublin. She realised that she was
in one of the nineteenth century European concession areas, and for a
moment history leaned over the buildings to embrace her, and the world
was a small place. These were old British mansions converted into shops
or civic offices or accommodation. With her phone, she took a picture of
washing-lines strung between Roman columns.
    A crowd was gathered in front of a mad-looking building that seemed
covered in plaster tendrils inlaid with broken pottery. The street walls and
parts of the building itself were inlaid with headless Buddha statuettes,
cracked vases and ceramic animal figures. She took pictures up close and
from across the road, imagining it through the eyes of Sandra, when she
would show her the photos later. Better hurry though, just over fifteen
minutes left to get back to the Costa Coffee. She left the scrum of jostling
tourists outside the museum, the honking traffic edging around them,
and ambled onward to complete a rectangle route back to the Costa shop.
More nineteenth century buildings, with some recent ones of the same
size in between, and directly behind them an incredible skyscraper being
built.
    Something was happening down a side street. Perhaps another market.
Monica went to have a quick look. There was a circular park bounded by
low railings, which contained an area of trees and a communal exercise
area. On the railings, long red banners were covered by lines of white A4
paper, and above the railings, wires had been stretched on which more
sheets of paper hung like washing, moving gently in the faint breeze.

Tables in front of these banners were also covered by piles of paper and stacks of thick clipboard folders. There was only a loose gathering of people, wandering around, looking with vague interest at the lines of paper. It was hard to say who was manning the tables, and who were the passers by. No money was being transacted. No one seemed in any hurry. It was impossible to guess what was going on. Mostly the sheets seemed to contain lists of some sort, written in Chinese. Monica focussed on one that had a bit of colour: it was a snapshot of a young woman. Other sheets also contained photographs of young men or women. Like the fliers you saw at home occasionally, in shop windows or taped to poles, of people who had gone missing. But there were thousands of sheets here.

She inhaled sharply, put her fingers to her mouth, and looked more keenly at the people sauntering around. They seemed resigned, or sanguine in some way. Not hopeful, but not sad either. And everyone was middle aged. She realised with a shock that they were parents, and the snapshots were of their adult children. Immediately the rumours of detentions, of false trials and imprisonments, of executions and organ harvesting, became darkly close. Monica backed stiffly away from the lines of paper. Everyone was sneaking little looks at her, all trying to seem unconcerned. A few older men stared openly. One even nodded and smiled. What the hell could he be smiling about?

Monica reached into her knapsack, put on the glasses.

'Hello? Peter? Peter?'

*'Amm, hello Moan-y-ca. Aww you back to coffee place?'*

'No. No, I'm not. Peter, can I take a photo here, and you tell me what's going on. As long as it doesn't get you into trouble, ok? I mean, I hope this isn't a problem.'

*'Ok. Let us see.'*

'Ok.'

Monica faced the lines of paper, the folders on the table. *SHICK*. She waited, acutely aware of the gazes upon her. What would happen if she got involved in political activities? Would she be imprisoned maybe? She tried not to look concerned, to be a carefree tourist. She turned, began to move nonchalantly away from the gaze of these stoical-looking men and women.

*'Aww, Mona-y-ca, this is organization of the parents, help young person find husband or wife.'*

'Oh. You mean... missing husband or wife?'

*'Missing? No no, find woman to marry. Or man.'*

'What, like a match-making thing?'

'*Match-making? Ooh, I don't know. Is like dating website. But they are old people, so no use computer. Only help son or daughter get marry.*'

'Ooh but that's lovely! That's so cool. So these are parents of people looking for a date?'

Now Monica wanted to go back, see what she could make out on the sheets of paper. Everyone seemed to be smiling now. Monica was smiling too, loose limbed and relaxed. An old man extended his arm invitingly toward a fluttering sheet. He was with another man, and the pair laughed, hunching their shoulders like two naughty children.

'*Yes, parents. Parents have plenty time, working person no time. Cannot meet girlfriend.*'

'So Peter, do your parents do this for you?'

'*Ahaa! Noo. Parents too poor, work hard everyday. They cannot come city. Is inconvenient travel from village cause road is too bad. In fact, they never go city all life. Not even Hezuo.*'

'Oh. I see.'

'*Also they want wife of Gansu. I want Gansu wife too. Wife from city no interest Gansu Province.*'

'Will you go back?'

'*Go back?*'

'To Gansu.'

'*Yes. Will get married Gansu. Live there.*'

'But isn't life good here?'

'*Life good here. But life better in Gansu. Especially if money. When get broadband link in Gansu, I will go.*'

'Wow. But just tell me, I'm interested – what's better in Gansu? For instance.'

'*Ooh. Food better. Air better. Can talk with persons after work. No so rush. Here all rush, all work. No time talk.*'

'But you came here yourself, right?'

'*Yes I came. For education. Then work.*'

'And is it as modern in Gansu as it is here?'

'*No, is not mo—*'

Monica waited. But the vitality of the earphones had disappeared. She took off the glasses, examined them, put them on again. Waited. There was no sound. She tried again. Nothing. Probably they needed to go in the case. She put the glasses in the case and started out at fast pace for the Costa coffee shop. When she was almost there, she put the glasses on again. There was no sound.

A sense of panicked reality came to her. As if the whole thing had been a dream. She had a moment of prickly heat, put the glasses back in the case, entered the department store with the Costa sign, and took the escalator to the first floor. It was just like any coffee shop anywhere in the world. They had exactly the same biscuits as they had in Dublin. She stood at the counter, looking for a man who looked like a boss. It was all too quick. Maybe he wasn't here yet, or this was the wrong place, or there never was a boss. How would she know him anyway? She was peering anxiously around when a small man in a white too-bulky shirt, dark blue trousers and black office shoes came almost beside her. She hadn't noticed him at all. He came up to her, bowed and grimaced, but didn't say anything. He wore thick glasses and his face was puffy and glistening with sweat. His hands twisted nervously together. He bowed again, and made a hopeful gesture, as if putting on glasses. Monica stared.

'Are you the boss? Yes?'

His grimace increased. He hunched his shoulders, bobbed his head. Cast a longing glance to where he had been sitting, but did not invite Monica to sit. Monica felt that every eye in the place was on them. He made the putting-on glasses gesture again. Said something in Chinese, looking hopefully at her. Monica made an elaborate shrug, shook her head sadly. She took out the aluminium case, and was about to hand it over. But she opened the case, put on the glasses one more time. It was no use. The magic was gone. They were just a pair of awkward, clunky glasses. She handed them over. Boss man seemed to relax. He clipped them into their case, and seemed all set to hurry off. But he turned abruptly, and digging in his pocket, he pulled out a wallet, hastily pulled a fifty yuan note from it. Bowing repeatedly, he offered it to Monica with both hands.

'No no,' she said, waving him off. 'Way too much.'

He insisted, bobbing his head and catching her arm. He thrust the note almost frantically into her hand, and as she tried to coax him toward the cash register to get change, he made a final bow and hurried off. She was alone in the middle of the floor.

The adventure was over. She looked around, but yes, it was over.

She stiffened, like a cat that has slipped, and began to make her way self-consciously out of the shop. At least she had enough money.

But what about the jacket? Apprehension quickened her step. There was lots of time before five o'clock. Only she better not get lost, there were no Chinese sunglasses to help her now. She wandered into the street, looked for landmarks. Better not be too early either. Four blocks from here, wasn't it? Or was it three? Oh dear. She should have written it down. Probably she would recognise the turn. She began to make her way along, and remembered with a little heart-thump that she had promised to show him what the jacket looked like on her. Little as she had trusted the glasses, or Peter, she missed them now. She missed the sense of back-up, the help with language, even the interaction for heaven's sake. It had been like having a friend. Imagine having a friend who was always in your head when you needed them. Maybe that's what Peter was planning. Rent a friend. You could be anywhere on earth, and have a friend from Gansu in your head. Monica had a brief sensation of flying, of looking down on clouds. It was so simple when you thought about it. But impossible now.

# III

# The Haggard

I was thinking would I chance going toward the town, when in comes this auld hag, not a tooth in her head and a huge hape a kids, sayin to me was there any hope of a bite of food to feed their hungry mouths, they hadn't a thing in the house since the father died and the mother went off to England. Bedamn, says I, I've nothing myself except stale bread that I takes off a the crows down alongside Coffey's public house. I gives it a wipe along the wall, says I, to put a bit of a taste in it, and you're welcome to a bit for yourself and the childer, I've a months supply saved up in the Nitrate sack beyant in the corner. Ah, says she, sure a man like yourself up here all on his own must have the finest of feedin what with all the money I have and she winkin at me.

I seen her givin one of the kids a poke and it starts bawlin and soon the whole hape of them is bawlin and some of them has red hair and more of them has black and more again is blond and at least five or six of every type and shape there is and still th'auld one is croakin at me what about a bit of money and still more comin in the door and not one of them over the age of seven, all screamin and bawlin and pullin flakes of distemper off a the wall and pullin the papers out of the good chair and they roarin at th'auld hag they were promised sweets. Says she what you need is a good woman and hadn't she her hand inside in me pocket. But divil the pocket there is in there only the family jewels swingin free and by Jaz doesn't she grab a hauld of them and she probably thinking twas a pigskin purse full of money. Well I left a squawk outa me that'd strip a heron but you'd hardly hear it inside in the din that was there that evening and I busted out through the back door into the haggard and away with me through the nettles and briars and over the ditch and away through the slurry tank where the top fell in long ago and down with me into the septic pit and up the other side through the furze bushes where the Conroys have generations of cars stored and down along the face of the quarry and into the swamp, until th'auld hag left go of the family jewels and she sank into the swamp and the children caught up with her and they roarin for sweets.

I staggered back round to the front and opened up the barbed wire gate that I have and up to the front door that hasn't been used in fifty years and sure it came away in me hands. Nothing would do then only for the rain to start tumbling down, so inside I went still holdin the door which I pushed back into the opening. T'was doin this I noticed that there was four legs on the door and sure the way I was after putting it in now weren't the four legs stickin out into the outside and maybe take the eye outa someone's head although no-one had called to the house since I could remember. By Jaz, says I, that's horrid like the table that went missing around the time of the mother's funeral. All the same, says I, better put the legs stickin inwards the way they were, and as soon as I did that didn't I find I was outside again and the rain pourin down on top of me. What harm, says I, isn't the roof inside lakin just as bad sure isn't it nearly all the wan.

So there I was standin in the downpour and along comes a lough of big lumpy men through the bushes. By Jaz, says I, maybe tis the county council come to repair the road, sure the main reason I never go anywhere is unless there's no rain for three weeks a fella couldn't tell if he was on the road or not, only that if you went in over your waist in water you were probably strayed off of it. So now seein these lumpy lads and they batin their way through the bushes, I left a roar at them and finally they espied the house about ten feet away and up they come towards me. Well, says I, is there an election on? Damn the election says the big fella and he catches me by the tip of the ear and drags me inside capsizing the table down flat with one kick. We want your money now says he and be quick about it. Yerra help yourself to all you see, says I, sure I've no need of money here and wouldn't know it if I saw it. Less of the saucy stuff says he and if you don't give up the money right quick you'll be the sorry man. How would that be, says I, curious to know what type of conditions might be worse than the current state of affairs. The other lumpy lads were all for batin me straight away, but after my adventure through the haggard and the slurry and the septic and the swamp, the state of me wasn't the best for layin hands upon, which was why the big fella had caught me by the tip of the ear where the rain probably exposed a bit of skin. They had an auld rope with them and they tied me onto the good chair and they all lookin round for something to bate me with but sure everything was burnt long ago and they could find nothing only maybe to pull apart the good chair and sure wasn't I tied to that. Where's your money roars the big fella or be Jaz I'll... I'll... And he seemed to lose the jist of it. Go on, says I. Have you that pliers Mikey, roars he at one of the lumpy lads, and they all still lookin around for money or something to bate me with. Up comes a lad with a rusty auld pliers and gives it to the big fella. I'll pull every tooth of your head he roars. Oh the blessins of God on you sir, says I. Now if he said he was going to pull the hairs outa me nose he'd have had the heart crossways in me, cos many's the time I hurt myself sore pullin a block out of the nose and the hairs hangin dearly onto it, but when he came on about the teeth I opened up the gob straightaway so'as he'd get a good swing at the three teeth remaining and they the biggest menace ever, paining me day and night and stopping me chewin the bread, only to soak it in rain and suck at it.

As soon as I opened the gob, they wrinkled their noses and screwed up their eyes and turned anyplace else. One of them says maybe we'll pull the fingernails out of him and that'll get him going. Right you are says the big boy, and they grab a hauld of me hand and they squintin at it and turning it around, sayin there doesn't seem to be any nails on it. Sure one time I used enjoy biting me nails when I had teeth in me head and nails in me fingers, but once th'auld badger ran off with the shovel there was no way of getting the spud only goin at it with the hands and now all the spuds is dug by hand and nary a nail there is on any finger only callouses thick as the sole of a boot. That one's the thumb says wan'a the lumpy lads, no this one's the thumb says another. But this here's the palm says the first. No it isn't says another, is that his left hand or his right hand? One of the lumpy lads takes a look at me from the side and I suppose what with me trousers being on back to front on account of excessive holes in the knees, and the well'tins on the wrong feet cos of a terrible hurry on me back in 1975, and myself looking back over me shoulder to see what were they at, the fella became horrible confused as to what way round was I at all and sure the poor man fell down in a fit thinking they'd broke me in half putting me into the chair.

What we'll do now, says the big lad, is we'll pull the toenails off of him instead. D'ye hear that, he roars at me, where's the money or we'll pull the toenails off of ya. Small harm to you, says I, and since you've a pliers maybe you'll be able for it and maybe you won't, but myself I haven't been able to get the well'tins off since 1976. Up to that point I used change th'auld sox fairly regular every New Years day. There'd be nothing left of the old ones only the rim and I used pull th'auld rim up the leg to make way for the new sock, then back on with the well'tins pretty quick before I'd suffocate. Only in 1975, what with the thickness of the smell, I closed my eyes and didn't I put the well'tins back on the wrong feet. The following year, I remember it well cos t'was twenty years after the mother died of bad luck from a half-choked magpie, I couldn't get the well'tins off for love nor money neither of which I had, and that was the end of the fresh sox. Be Jaz, says wan of the lumpy men, maybe tis inside in the boots he has the money. If it isn't, says the big lad, out with the toenails.

Well, half the lumpy men laid hold of me arms and the rest laid hold of me boots and they went to heaving like a tug-of-war. After half an hour, when I was about six inches longer and nothing else happening, didn't one lad have the idea of heating the boots with fag lighters. They went at it again, yer man full pelt with the fag lighters, and next thing there's a sucking noise and a blue flame and the boots went flyin and the whole place full of the most atrocious stink ever smelt above ground. Bedad tis worse than '75, says I to myself, keeping th'auld gob shut for fear of asphyxiation, while the lumpy men were going purple and boggle-eyed with the stench such as no septic tank ever left loose. In the hope of fresh air, I put the head down and busted out the back door into the haggard and away with me barefoot through the nettles and briars and over the ditch and away through the slurry tank where the top fell in long ago and down with me into the septic pit and up the other side through the furze bushes where the Conroys have generations of cars stored and down along the face of the quarry and into the swamp where there was an awful splashin and screechin and roarin of children and next thing I see th'auld hag and she says There he is and the whole she-bang takes up after me and what with the lumpy men comin behind me I says to myself tis into the thicket for me and I running along with stones and sticks and bits of rusty ganvanizing sticking to my feet and sure I thought the smell was catching up on me but wasn't it comin from the feet themselves. So for the pure love of life I bate through the whitethorn thicket and off into th'auld stream and after enough of stones had stuck to me feet and fell off again weren't me feet back to their normal size at least, so I circled back to th'auld cabin to see could I get the well'tins on again right way round this time and maybe risk a breath of air.

I sat down in the chair, trying to pull on the well'tins without taking a breath and listening to the faraway sounds of roarin and screechin in the thicket beyant, and bedad no sooner had I the well'tins on and takin a good suck of air and realizin I was after missing the chance of putting on fresh sox if only I had them, when in over the flattened table comes a dapper man with a beady eye and says he to me do you not know me, I'm your long gone father, back from America. Begob, says I, you're hauldin it better than I am, cos the only hair I have on me head is in me nose and ears and sure yourself has hardly even a bit of grey. Ah tis the good life, says he, and hardly was the words out of his mouth than in through the back door is another beady-eyed fella who says I'm yer long gone father, and while them two is takin at one another, in comes another wan and another wan, and next thing they seemed to be comin in the window or down the chimney cos I could hardly see the wall with them and each of them waltzin in and out whispering is there the loan of a few bob to get a business going or thaw out a bank account or pay a ransom and other such things and they had the eyes going round in me head and me with no father at all my whole life and now about fifteen of them of all ages. I was wonderin did the mother fall asleep naked in the dressin room of a hurling match or how did it happen I'd so many fathers, and distraught I was at the memory of years toiling away with no father at all and the mother sitting gobsmacked inside in the chimney, probably broken hearted at the fifteen men who loved her and went to America.

Perplexed with the sorrow of it all, I espied a hole in the wall that I thought I might lep through to get outside but sure wasn't it only a patch of rising damp and didn't I dive headlong at the wall instead of through a hole, poleaxing meself entirely and falling in a fuddle on the floor. Give him air says wan. Give him water says another. Give him room says a third. Give him here says the last. I woke up in a coma, across the back of this lad and he batin across the haggard and away through the nettles and briars and over the ditch and off through the slurry tank where the top fell in long ago and down into the septic pit and up the other side through the furze bushes where the Conroys have generations of cars stored and down along the face of the quarry and through the swamp and into the thicket where there was raucous row and ruille-buille and th'auld hag with a well-aimed stick brings down the father that's carryin me and the rest of them following is tearin at each other and batin into the lumpy men with the horde of children entangled in their legs and biting their knees. Luckily for me being on the ground and the same colour as the mud itself they couldn't see me at all, so away with me slitherin back the way I came, over ditches and dykes and through all manner of filth and rot towards the haven of my quiet little cabin.

Hardly had I clawed up onto th'auld good chair than in comes a Vision of Lovely Maidenhood, like the picture of the Blessed Virgin Mary that's underneath the ivy above the fireplace, except she didn't seem to be wearing a skirt at all only a thing like a wide belt and she had heels on her shoes the length of a garden fork and what blouse she had on was belongint I suppose to her baby sister and could in no way disguise her ability to suckle youngsters, all a'this with lips of Massey-Ferguson red set above by the widest innocent eyes that ever surveyed the townland.

Are'u lost, says I, and she standin there sort of squirmin' like, one knee in front, then the other, her hands on her waist, and every time she did it I got a different view of th'auld paps and by Jaz, says I to myself, if she keeps that up she's going to bursht out of her clothes entirely and begob I was thinking the crotch of me trousers was in similar peril of burshtin asunder even though the trousers was still on backwards due to excessive holes in the knees.

Next thing a hiss through the window says Get the clothes off of him.

Well, that sets the Vision all in a flap and she looking at me like a rearing horse might look at a dog and says she would you not like a shower and get out of them durty clothes. Divil the shower here, says I, only full scale rain coming in through yonder gap in the roof, where I does have ablutions betimes while I'm asleep. Get the clothes off of him, says the hiss in the window, and th'auld Vision starts wringin her hands and eventually she shuts her eyes and grabs hauld of th'auld jacket on me that used be the Sunday best 'till it went in flitters up to the shoulder and is only steadied now with dried mud such that if I were to be caught out in a high wind it'd nearly be the death of me with the flappin it does be at. Fair play to the Vision she must have had some experience cos she got th'auld jacket over my head and flung it into the corner where it broke in four halves.

Oh Jeez tonight, says she, he has a woolly jumper on. Divil the jumper, says I, only a string vest I found abroad in the stream and didn't the hairs on me chest and back grow straight through it. And he has about a half dozen of necklaces on, says she to the window. God bless you, says I, they're the frames of the previous vests I had on.

Well at that th'auld Vision made straight for a hole in the back wall, and a screech comes in the window Get back there you lazy slut if we don't get a picture of him naked we can't say he took-at you. And in through the window comes an auld crone in a doctor's coat and she an orange face on a pullet's neck and pinky lips puckered like a duck's beak. T'wasn't fear of her camera sent me flying through the hole in the wall after the Vision.

Not far I got however, cos there was the Vision sunk to her ankles and she blockin the gap in the haggard wall, and up through the nettles and briars comes the fathers and the lumpy men and the hag with her screechin children. Oh goodnight, says I, and turned for the bog, where people often went in and were never seen again. But barely was I up on the collapsed mound that used to be the pigsty when I seen a tidy fella comin in through the barbed wire gate, careful-like, and he in a fancy suit and half-spectacles and carrying a tidy little suitcase. I knew he must be the most awful ever cos the whole lot of them behind me went suddenly quiet save the odd clap agin a child's ear. See gee shees wrong a hane, says th'auld hag and she putting a full set of teeth into her mouth, and says a lumpy lad to one of the fathers Is that you, Father Considine. Tis Sergeant, whispers he, is that Doctor O'Donohue there?

Up comes the tidy man to the top of the pigsty and the whole lot of them gathers around. Says he would you be the man known as Dan Dinny Anne. Deed I am, says I. At long last, says he. If you will sign here I will give you a cheque for one hundred and forty seven million euros, for you have won the lottery. Well, says I, that is an amazing fact. It is, says he. Specially, says I, since I never bought a ticket. What's that, says he, and he cross as a wasp. Aren't you Dan Dinny Anne, known as Dan Dinny Anne the Bog? No says I, I'm Dan Dinny Anne the Haggard, Dan Dinny Anne the Bog lives beyant in the swamp.

Well by Jaz I was thrun upside down by the stampede. When t'was over there was only myself and the tidy man left above on the pigsty. He looked down at me a moment and with a small sigh starts pickin his way back to the barbed wire gate. Good riddance, says I, and went in to sit on the good chair and damn the goin to town.

# Boxes

The first I heard of these food boxes was Ma and Da talking in the kitchen.

'It must be the Chinese,' Da said. 'Or the Americans maybe. It can't be the Russians – they've hardly enough for themselves, let alone this kind of craic.'

Ma shook her head all the time Da was talking. It didn't mean she disagreed with him, that's just what she always did.

She said: 'First place it happened was LA, remember? And New York, at the same time. It said it on the news.'

'That's only when it got to be a big deal. It happened first in what's-its-name, that place in India.'

'Delhi.'

'That's right. And someplace in Indonesia.'

They were standing at the kitchen table, leaning over the newspaper. A photograph on the front page showed a couple of men in suits and a policeman, posing beside some white cardboard boxes. There were lots of other boxes away behind them, and people milling around as if there was a sale or a football match. The headline said: Food Boxes Left on City Streets.

Ma said: 'But what's the point of it? Why're they dumping food round the place?'

'They're not dumping it, they're giving it.'

'That's what people said with those poor places, but we're hardly starving, now, are we? And New York's hardly starving.'

Da furrowed his brow. I didn't get involved. I was only hanging around to see if I could get Da on his own, see if he'd sneak me the keys of Ma's Clio to bring Jenny to her music class.

Ma turned her head. 'Peter, haven't you study to do?'

'Aah…yeah, I'm gonna do that after dinner. I kinda need a bit of a break.'

'A break? Look, if you don't start taking it seriously pretty soon, we'll have to get you into–'

'Oh Ma, don't start! You're at me every single day, and I told you already, I'm *on* it.'

'What do *you* think?' she says to Da.

'Hmm…?' He was frowning at the newspaper, rubbing his lip with his knuckle.

'Well?'

'Just a minute…'

'Oh for heavens sake!'

I got fed-up hearing about these food boxes. There seemed to be nothing else on the news. Ok, it was weird that no one could say who was doing it, or where all the food was coming from. But they were making such a deal about it, and on the TV you couldn't know what was true or what was spin. But then Jenny got interested, so I got interested.

I still couldn't believe my luck that Jenny and I were… well, let's call it an alliance. I hadn't had the guts yet to actually try and snog her. She might take that badly, and I'd lose this special situation I'd fluked. She'd only been in our school since Transition Year, and since then I'd basically had to train my mouth to stay closed so my tongue wouldn't hang out. But she was no dummy, and she definitely had her own way of doing things. She didn't wear make-up or go around in a bunch like most girls did. She didn't come to football matches either, which was bad news for me, and most of all she didn't drink alcohol, so there was no way I could ever get her relaxed. We'd probably never even have got talking only she heard me answer in class that I'd actually read a book, The Catcher in the Rye, cos of the time I'd busted my ankle in Junior Cup, and Da, God bless him, brought me in this ancient book instead of buying what I asked for. She was the only person I'd ever met who'd been on a demonstration – against climate change, with her parents. She wouldn't eat battery chicken, or use soya from Brazil, or sit at a mahogany desk. She wouldn't eat fast food at all, and even healthy food had to be organic or free range or have never been on a plane, or something like that.

One time, at the end of Fifth Year, she roped me into volunteering in an old-folks home. It was just a one-day thing, and I felt good about it afterwards, even though the wrinklies made me kind of squirmy, and

Dean and the other lads on the football team gave me an unmerciful slagging. I preferred the time we helped out on a charity fundraiser on phones. That was a bit of craic, about a dozen of us in a big room in a real office, like we were real workers. It was for flood disaster relief in someplace I never even heard of, but we raised more than the amount they targeted and Jenny put her arms around my neck and hugged me and ooooh boooy I was the king of the world. The lads could slag me all they liked.

And then these boxes came along.

Ma and Da went to see the food boxes in the car park of the shopping centre. Afterwards, Ma was saying: 'If the Food Safety Authority says it's perfectly good, then it should be put to good use.'

'Yeah, but they can only sample so much,' Da said.

'It's stupid to just waste it, and – my God – the amount of it. I mean, something has to be done with it.'

'It's not as simple as that. You can't just drop your way of life.'

'But something has to be done with it.'

'I know that.'

'Well, we should use it. Everyone's using it. The whole world's using it.'

'Yeah? Well, you see all those farmers and truckers picketing in the city centre? D'you want to go in and explain that to them?'

'But farmers can use this food like anybody else.'

'Oh my God, that's not the point!'

'But why should we pay five hundred euros a month for food when it's free in the city?'

'Because nobody understands it! We have to have dependable food.'

'But this has been coming regularly for weeks now.'

'Mary – Mary, listen to me. This is totally crazy, this food. No one knows where it's coming from, ok? No one can explain it in the slightest. D'you not get that? We cannot depend on it.'

Ma stuck her lip out, the way she does when you're not going to win the argument.

'Father Moran says it's happened before. It says it in the bible.'

Da put his hands over his face. 'And some guy on the telly says it's the damn aliens. One's as likely as the other.'

'Look, millions of people have believed in this kind of thing for thousands of years. They can't all be wrong. Father Moran says we've been living in a time of science, and that's all over now.'

Da stared at her. I chewed my sandwich as quietly as I could, looked from one to the other. Ma had a weird look in her eyes.

She became suddenly all practical, marching round the kitchen putting mugs and plates away: 'If this is a gift from above, we should take it and give thanks.'

'Mary, do you not see what's happening? *No-one* has claimed responsibility for all this, and now every religious crackpot and sci-fi nut on the planet is getting in on it. Let one of them show me how it's happening or how it's going to last and I'll use it. Until then we're not touching it!'

'We're the only people in the estate not using it. Everyone else is using it.'

'No they're not,' I said. My parents looked at me like they'd forgotten I existed. 'Jenny's family aren't using it.'

Ma wrinkled her nose. 'They're... odd.'

A couple of days later, on my way to her house, I met Jenny stamping along the road. Her eyes were red. She was so mad she couldn't speak.

'I... don't... be*lieve* it...' she sobbed.

I had never seen her like that before, and it wrenched my guts. I put my arm around her and didn't even think how nice it felt. She told me her parents were using the food parcels.

'They're... *stupid!*'

We sat on the wall beside Macari's Pizza, my arm still tentatively on her shoulders while she dragged chunks of air up her nose. I didn't know what to say.

She turned to me. 'Are your parents using them?'

'No...'

'Can I come and stay with you then?'

I puffed-up.

'Course you can. Long as you like.'

Ma was pretty dubious about it, but she phoned Jenny's parents and they were totally laid back. I couldn't believe it. Normally Jenny's parents wanted everything to be discussed to death with all sides being consulted and everyone having their say, and then her Da pontificating for about half an hour. The result of all this openness and fairness was that I was never even allowed in her bedroom. And now she could stay in my house.

We had dinner of steak and kidney pie from a tin. Jenny poked at it glumly and concentrated on the vegetables. Ma said there was no fresh meat in Super Valu and Denihy's butchers had closed. For the previous couple of weeks, there'd been heaps of meat everyplace, quarter price, to try to get people to buy some. Ma said she'd have stocked the freezer if she'd known they were going to throw it out.

Da quizzed Jenny about why her parents were using the food parcels.

Jenny said: 'Yesterday they tried a tart, just to see. They got it from Mrs McMahon across the road. Today they got their own boxes.'

'What kind of things were in them?' Da asked.

'Just like we buy. Everything.'

'And where did they get these boxes?'

'There were two of them outside our house this morning.'

Ma and Da exchanged glances – Ma triumphant, Da suspicious.

'What was the food like?' Ma asked.

Jenny looked at her carefully. 'They said it was excellent.'

Ma smiled smugly. 'But you didn't want it.'

Jenny shook her head. She watched Ma for a moment, and cast a dark look at me.

School was cancelled temporarily. Probably because there were arguments going on all over the country, like there was in my house. Not in Jenny's house. Her parents were in great humour. I'd never seen them so relaxed. They didn't seem to mind Jenny staying with us for as long as she liked. We collected another bag of Jenny's things and they waved us off. They were having drinks and listening to music, even though it was twelve in the morning on a Wednesday. Jenny marched back to our house almost at running speed. I scuttled along beside her and kept my mouth shut.

The day after that, Da came home early from work. He's an insurance broker, with a pretty good setup of his own at the top of Henry Street. We met him driving in the gate, and he said he'd been the only one in there all week, no clients either. So he came home. I saw the look in his eye. He slammed the car door, headed for the house.

'Oh oh,' I said.

He and Ma had been arguing constantly, about food, about politics, about all sorts of things. Ma wasn't in the house, so he came back out and started tidying the garden, just for something to do. The whole neighbourhood was really quiet. There seemed to be very few cars driving around those days. The previous football training session, not one single other person turned up. Not even Dean. He didn't answer his phone, and he was never in his house anymore.

Jenny and I went up to the spare room where she'd been sleeping. We'd been contributing to an Internet forum, but it seemed to be going nowhere, just dwindling away. She had an idea about trying to get our friends together, start a local discussion group, and she wanted to make a list of things we needed to do. It was kind of exciting.

Ma called us down for dinner. She'd put together a fantastic-looking spread. Beef Wellington, croquettes, sweetcorn, baked fennel, gravy, and a trifle, plus a few bowls of some sort of stodge. After what we'd been eating the last week or so, this was a big improvement.

'Now,' she said, 'I'm giving you the choice. You can have all this,' she indicated the meal, 'from the food boxes. Or porridge,' she indicated the bowls of stodge, 'which is the last food I have left. There's nothing in the shops. Which would you like?'

At least she was pleasant about it. Normally, if she made a dinner like that, she expected you to say your appreciation out loud, at the beginning, in the middle, and at the end. Today she was totally relaxed.

Da was not relaxed. He stared at the food, frozen in mid expression. He put his elbows on the table, his head in his hands. Finally, with a long sigh, he sat up, lifted his knife and fork like they weighed a ton, and took a slice of Beef Wellington.

Jenny said: 'I'll have the porridge.'

You could see a knot of muscle in her jaw. She turned to me defiantly, but her lip quivered. 'What about you?'

'Oh…' I said.

Beef Wellington was my favourite. And we'd been having such crap food for ages. But I'd never been as close to anyone as I was to Jenny right then. We'd been having comforting cuddles in the spare room, and I had this delicious melting feeling of things to come.

'Sure,' I said. 'Jenny and me'll have the porridge.'

Ma murmured: 'Great,' and we all sat.

'I'll make some more,' Jenny said. 'D'you have any honey? Or nuts?'

Early next morning, Jenny tiptoed into my bedroom. I woke up with the usual early morning boner, and for a moment I thought my fantasies were coming true. But she was fully dressed, and her face was grim.

'Come on,' she said. 'Come and see this.'

I put my clothes on and we went downstairs and outside. Under the awning at the front door were two white boxes, just ordinary-looking cardboard, with little openings you could get your hands into for carrying. There was no writing on them.

'This is how it comes.'

We looked at the boxes in silence.

'The first time, in India, they had writing on them. No one even recognized the language. Now there's nothing.'

'How'd you know that?'

'The Internet.'

'And you believe that?'

Jenny shrugged.

We lifted the lid off one box. It had a carton that said Milk, another that said Orange Juice, four small tubs of yogurt, a small loaf of sliced white bread, a smaller one of brown bread, some fruit, and a baked pie in a foil dish. Everything had a beautiful picture on the front, to show what it looked like ready to eat. The other box had a container that said Pork Table Ready, and that was warm. There were packets of cheese, ham, olives, three types of nachos, muesli, tomatoes, a tub of sliced cooked potatoes, a bottle of wine, and a lot of other items. We didn't bother looking at the rest. It was just like stuff you'd see in a shop, except there was very little writing, mostly pictures. All the writing was in English.

'Your mum put out yesterday's used box, and this is here today.' Jenny fixed her eyes on mine. 'Promise me you'll never touch it. Not even one biteful.'

I shrugged. 'Sure.'

'We don't know where it comes from.'

I laughed. 'Jenny, we don't know where anything comes from.'

But she wasn't really listening. She said, 'We have to find some more food. We can't live on porridge.'

I was able to get petrol for free, because there was no one attending the Maxol station or the one at Chalke's, and all the pumps were left on. Soon there wasn't any petrol left. One morning, we woke up, there was no electricity. So there was no TV. There was no Internet. You'd think people would be upset, but everyone just sat in their houses, eating and drinking and talking. They were all in terrific form; big easy-going parties, get-togethers. People stayed over with each other, people who hardly knew each other. The parents who'd been gassing on about religion calmed down, and so did the conspiracy theorists and the sci-fi nutters and the activists. No one had any worries. The boxes were bringing food that was warm, so they didn't even have to cook it. Even Da was in party mood. He stopped putting on a tie in the mornings.

Jenny and I spent our days on bicycles. We were in a kind of private adventure together and I loved it. We found fields of carrots and turnips, less than a mile from our house, in a place I never knew existed. About thirty-five minutes cycle away, we found potatoes in a fenced-off section of a field, half-overgrown, which we could cook on Da's camping stove at home. This took all the pressure off us, because two shopping bags of potatoes brought home on the handlebars would do us for a week. We spent a lot of time letting animals out of sheds. Then we came across a house that had a proper vegetable garden. A lot of country houses seemed to be empty, but a woman lived in this one still. She said we were welcome to the vegetables.

'Why don't you use them yourself?' Jenny showed her two huge parsnips we had pulled, and some red stuff Jenny called chard.

The woman smiled. 'I'd have to go out and pick them, then clean them, chop them, and cook them. When I've got these.' She indicated a white box on the worktop inside the window. There was a carton of foil beside it, like you used to get in a takeaway. On the top was a lovely picture of mixed vegetables.

'They say the food's even better in the city,' she said.

Next time we called, the woman wasn't there.


We were coming home at dusk one day, just entering Sallywoods estate, and we saw a young guy walking in a peculiar way out the back way from the estate. This guy was about ten or eleven. Neither of us knew his name but I'd seen him before, cos he played for Milford GAA. He lived in Beechwood Crescent over in Beechwood estate. He was drifting along with a sort of carefree, jerky walk, through the grass area where they'd left a few big old trees. The grass hadn't been mown there for quite a while, and he had to lift or drag his legs through it, and step over an occasional bicycle or bin, as if he couldn't go around those things. We waved at him, and he waved vaguely back. He had a sort of dazed expression. He reached the main road and went off along the tarmac in the direction of the city. I realised finally what was strange about his walk: it was like he was trying to stop himself hurrying, like he was going down a steep slope, not a road that was level. I called to him. He turned partially, made another wave, and carried jerkily on. It was getting dark, and something about the way he was going gave us the idea that he wasn't coming back.

We found the house in Beechwood Crescent where that boy lived. Gallaghers was the name of the family. Their living room was full of people snuggled together on the carpet, chatting and eating, but none of the Gallaghers were there. We went to four houses, enquiring, before we found the father.

'But he's gone away off along the road,' Jenny told him.

'Ah, they all go someplace every day. He'll be back for bedtime.'

'But he didn't seem to know what he was doing.'

'The kids are fine. Let 'em enjoy themselves.'

They were having gin and tonics. Ice, and pre-sliced lemon, lime and cucumber came in the boxes.

I wanted to tell Ma and Dad about that boy.

We found Ma in a house in the street next to ours.

'The bigger cities have more things in them,' she said. I'm not sure what she meant by that. Her voice was slurred.

She wouldn't come home. I couldn't find Da anywhere.

We started trying to convince people not to eat food from the boxes. They all listened politely, smilingly, and you could see they wanted to be nice to us, but they couldn't seem to focus on what we were saying. When we asked them to stand up and come and get some real food, they shook their heads benignly and asked would we not like some lovely köfte or pilau or pizza, or a glass of wine. Jenny's mum was nowhere to be found. It was four in the morning before we gave up looking.

Jenny took to clutching onto the back of my hoody. I'm not sure whether she was afraid I'd eat something from a food box, or run away, or that something might happen to her, or to me. Anyway, we were together all the time except when we went to the toilet. We started sleeping in the same bed because no one seemed to mind whether we did or not, and it was a big comfort. The third night that we slept together, we had full sex and we didn't even think about contraception. It was warm and wet and sad and lovely. There was a tremendous sense of being part of something huge, and of loss too, and of what would never be. We kept saying all the things we were worried about, and then we would join together again as if that would make it better. We cried on each other, and fell asleep entwined.

One evening, after a day of foraging, we found an old-fashioned farm cottage that had the remains of a walled vegetable garden, chickens that went into a coop by themselves, and a spring of clean water. It would have been perfect if it had an orchard, but it had only one small apple tree with a single ripening apple. But there was something optimistic and cheerful about that. This place was about half-way to Nenagh, and it was getting late, so we had to decide quickly – go home or stay there. No one would miss us back in Sallywood, we kept mostly to ourselves those days, and anyhow, at that hour, we would be the only two not drunk. Clean water was the deciding factor – there was no tap water in our estate any more, and the stream at Groody and the Mulkear river stank something awful, so we were becoming a little smelly as well as thirsty. We had a good wash and filled our bottles. We went home next day, but returned the same evening to that farmhouse. It had a wood-burning range to cook on. We decided that we would try to maintain the garden. Maybe even keep a cow.

It was amazing how much work there was in just one tiny little piece of ground. We figured out that if we were to make a real go of it, we could only go home at weekends. But the first weekend we were too tired. There was too much to do, because of stray dogs and foxes harassing the chickens, even a pair of stray pigs that came and destroyed most of the garden in one evening, and of course weeds everywhere. We had to start thinking about repairing the garden walls, storing some vegetables before winter came, and getting wood or whatever we could find for the fire.

When we went back to the estate the next weekend, the doors were all open. All the rooms were quiet and empty. There was litter everywhere: shoes, spectacles, wallets, jackets, a half set of dentures. No people. There was no noise, no smoke anywhere that we could see, no engines. Only the soft movements of the wind, the lazy call of crows in the sycamores by the new school gym.

How this silence made me feel, I can't describe. It seemed at some level that I had been expecting it my whole life. But I had to stop myself thinking, make myself some-way cheerful, for Jenny's sake. She was stumbling around, mouth open, arms hanging. I let her wander, following right behind her for more than an hour, as she tried the homes of everyone she knew. We went down to Annacotty, to see if I could find Dean. There was no one there either. As we came out of Dean's estate, a yapping and barking commotion approached. It was a pack of assorted dogs, about twelve or more, almost every kind you could imagine – White Scotty, Labrador, Beagle, Dalmatian, Alsatian, a huge floppy Great Dane. It was funny in a cartoonish way. They were chasing three emaciated cows up the middle of the street.

I took Jenny's hand.

'Come on,' I said. 'Time to go home.'

# The Museum of Future Art

Wesley gets off the bus at Arthur's Quay, and stands undecided in the walkway. It will be an hour and ten minutes here before he can catch a bus to the nursing home. So should he tramp to the bus station? Or hang around here at the downtown stop. If only the employment scheme would release him an hour earlier, he'd be able to get a direct connection. Or if the next bus to Newport would come earlier, or if the nursing home was more convenient, or if a million things were not just exactly the way they are, he would not be standing here like the last of a species, flapping around uselessly, waiting to be extinct.

His mum had always been so much older than other parents, and odd with it, and forgetful. So when she first started wandering away, it had seemed nothing strange. One day he'd come home from school, and Mrs Clancy – their landlord and neighbour – was there grim-faced at the back door of the cottage. Mrs Clancy raised her finger, and pointed at Mum in the swampy pond halfway up the field. Had he not noticed anything lately – Mrs Clancy demanded. He'd had to wade in, lead Mum out by the hand. Her expression had been of something approaching joy, but she had looked at him peculiarly, and wasn't really pleased to see him. He received a stern visit from the Community Nurse. The Clancys came to talk to the nurse too. They shook their heads, scowled and muttered.

But none of them knew what Mum was like – a whole lifetime of wandering and forgetting. Whenever she got drunk or had a crisis, she would pull him to her, wet his face with her tears, caress him, make him promise never to leave her, to mind her and love her always. It was infuriating. Did she think he hadn't noticed how she managed to buy paints for him? Even when they couldn't afford food. Or how she had found money for art classes, when what she really wanted was for him to be an accountant or a doctor or an engineer – anything, in fact, that might save her from being a cleaner. Mum had no talent for making anything clean. She never remembered telling him, on his seventh birthday, how on a doorstep in England she had decided to grant life to a bump.

Jed Clancy found her wandering on the spoil heaps of the old slate quarry, a quarter of a mile up the hill. Another time, Gardaí had to bring her home from Ballina, four miles away. They found her sitting in a boat. That was during his Leaving Cert. The only thing he could think to do was lock her into the cottage during the day.

During his first term at the School of Art and Design, he had arrived home one evening to find a fire brigade at the cottage. She had put her still-running hairdryer into a drawer of papers, and almost burnt the cottage down. The next day, Doctor Rennison and a Garda came to see him. There was a big discussion in the hallway, with Mrs Clancy standing outside, arms folded. Mum could no longer be locked in the cottage, nor could she be allowed out on her own; they would take the matter to court if they had to.

So he had to stay home. He was given a carer's allowance, and a helper came every day to help him cope. Then a helper came twice a day, then three times. One morning, while he was painting in the bedroom, she set the kitchen cupboards on fire by putting a book in the toaster.

After the fire brigade left, Wesley gathered all his painting things, put them in a large cardboard box, and put that on top of the wardrobe in his bedroom. He took all the canvasses – thirty-seven of them – and posters, wooden sculptures, drawings, sketchpads; took them all up the field behind the pond, doused them in white spirits, and set fire to them. The paintings spat and fizzled, the flames roared transparent red and orange in the sunlight, the smoke was abundant and black and bad-smelling. He felt lighter afterwards. There was no longer any strain. If Jed and Mrs Clancy had not been peering over their wall fifty yards away, he might have fallen to his knees.

For a while they lived in a life without days and nights, only light and darkness. Eventually the Public Health nurse – a no-messing farmer woman who spoke in a puttering manner that made her speech seem even faster than it was – told Wesley candidly, in front of Marge the Latvian helper, that in her opinion he had not the ability to take care of himself, let alone an incontinent and totally dependent old woman. He had to stand there, eyes down, and endure that speech. Afterwards, they organized a nursing-home place for Mum. In Newport, two bus journeys away.

On a sunny-bright afternoon sometime last winter, he and Mum went to Newport in a taxi, with a small suitcase in which there was still some room after all her things were inside. Mum smiled vaguely during this journey. She looked around, attracted by the movement. She would not take his hand anymore. When they reached the home, she walked beside him in a frail but carefree way, curious about everything, just like she was at home, the bulge of her incontinence pad ridiculously obvious on her bony frame. The admission process was without interrogation at least, and everyone was unbearably cheerful. He stayed late, long after she finished picking at her dinner. He did not eat or drink himself, although they offered him tea and he was starving. When he finally found himself outside, the sun had gone down, and he had not the slightest idea in which direction home lay.

It was past midnight by the time he reached the cottage. He was humped, shaking with cold, foot blistered. He hobbled to his bedroom, took the big cardboard box with his painting things from on top of the wardrobe. It contained half-cleaned brushes, scum-dried pots, paint knives, balls of multi-coloured rags, cap-less tubes of paint that had been abandoned on the turn of an instant. He raised the box over his head, tipped the contents to the floor. Kicked everything around, stamped on anything that could be broken or squashed or spilt. The reek of turpentine was powerful and mellow, the smears of paint on the mouldy carpet were a viscera of pain and memory. There was no one to see him, so he threw himself down, flailed around, covered himself in fluffy gritty multi-coloured ooze, until finally he was exhausted.

An employment scheme, otherwise they would stop his Benefit. 8.30am to 12.30 every day – picking litter, watering flowers, weeding graveyards. With big-bellied, split-nailed men who made sport of his clothes and mannerisms. At break time on the first day, he sat in the tearoom of the community hall, curled like a periwinkle inside his big black coat. A stumpy, bow-legged man came and stood at his shoulder. A stink of old cigarettes flavoured the air. He pointed at the three spoons in Wesley's cup, which was Wesley's way of counting the sugars.

'Tell me,' he said to Wesley, 'what cloud are y'on at all?'

Wesley's way of holding a shovel set them coughing and hocking with laughter. One of them had to turn away, put his hands on his knees, and spit a wobbling blob of mucus into the gutter before he could continue laughing. Anytime Wesley spoke, they laughed, or appealed to one another as to what type of specimen they had here at all.

Back in the privacy of the musty cottage, Wesley waved his arms.
'What the hell do they know? Fuck them!'

He stamped around the empty house, cast himself onto his bed. Wept. He would have stayed like that for hours, only it took so long to visit Mum. So he rose, dried his face in the stiffened T-shirt on the back of the door, and set off.

'Mum, it's me. Wesley. You know…? Wesley.'

Every day he would sit in front of her, leaning forward, trying to catch her eye. Her expression of wonderment would scan past his face like he was part of the air. When he touched her hand, she drew it away in surprise, as she might from cobwebs. The staff zoomed past in their white shoes.

One day, one of the Irish attendants said – right in front of Mum – 'Not much point coming every day, eh? You're not doing anyone any favours.'

Now here he is, trudging toward the bus station. It seems he has been trudging for years. Dearly he would like to make a roar, straight up into the air, but there are people around. He turns a corner, into a street of vacant modern buildings, old redbrick townhouses. A mist falls from grey sky. He plods onward.

There is a sign by some metal railings:

THE MUSEUM OF FUTURE ART
TUE TO SAT 10.00 TO 17.00

Wesley stares at the sign, plods onward.

What the hell did that say?

He stops, takes a quick peep around. No one is looking, so he whirls violently. He is perfectly entitled to whirl violently. He strides – *strides* – back toward the sign.

But a woman is coming, laden with shopping bags. Wesley slows, averts his face. He moves to the kerbside as if he has remembered something of great importance. Makes a nervy search of his pockets, a patting of his coat. The woman's face is weary. She ambles past with the roly-poly motion of bad hips, barely glancing at him. Wesley maintains his frown, concentrates on the important search of his pockets. Takes a peek at the woman's receding back, then a quick scan to see if anyone else is around, and goes to confront the sign.

There it is.

He extends his arms in appeal. No one is within sight, so he flails around, appealing for witnesses to this ridiculous development. Finally his indignation is spent. He stamps off toward the bus station.

Museum of Future Art how are ya!

But he can no longer visualise himself moving along the street. He can no longer imagine how his thoughts would seem if he were to observe himself from a vantage point. Now there is only himself, in an empty street.

He checks the time. Nearly an hour until the bus, and it's only a few minutes walk to the station. If there had not been enough time, he would have been compelled to hurry onward. But the way it is, he has to make a decision. He bites his knuckle. Begins another search of his pockets. Finds a pellet of fluffy paper that was once a tissue. He examines it, uncurls it. Flings it angrily, guiltily, to the gutter. To hell with their litter laws. Hasn't he been sweeping streets and picking up litter for months? Is he not due some respect for that at least? He swells with feelings that he cannot name. Turns back towards that sign.

He knows what will happen: a haughty ticket-person with arched brows, laying down the law to this scrawny degenerate in off the street. No. Not this time. This time he, Wesley, will take a stand. This time he will be unstoppable. *Future Art* indeed! They have it all figured out, everything catalogued and preserved. It is all over, and he has not even begun.

Wesley walks faster.

He can see the headline:

## LONE MAN DEFENDS THE RIGHT TO MAKE ART

But of course no one will applaud. No one will care. His breathing becomes rapid. He arrives at the steps, trembling. Takes a last look around. No one is near, so he leaps up the steps, distaining even to glance at the sign.

The door is of plate glass. Not a door you can swish through. A big, complicated handle. More like the entrance to a bank: two doors in line, with an area in between. Wesley takes a deep breath, grips the handle. But... he cannot budge it. Is the place closed? He steps back, checks his phone. 16.30. It should be open, no question. He searches all around, gazes up at the impassive façade. Finally, in his own crow-like reflection, there seems something written in transfers on the door:

## Insert €4 in slot

What the hell?

Wesley examines the handle. Is that a slot for money? Like a public toilet. He puts his face to the glass, cups his hands around his eyes. There is light inside, but no sign of movement. Is it open at all? But €4! – he rummages miserably in his trouser pocket. His hand comes out with… €5.26. If he spends €4, he will only have €1.26 left. Each bus-ride costs €1.50; there is one more to the nursing home and two on the way back. If he can't get a refund, it will mean a hell of a lot of walking. No, it is too hard. And the time is past. His choice was made long ago: *ART OR LIFE* – that's what it says on the T-shirt hanging on the back of his bedroom door.

Even if he does make a stand, why would anyone care? Why would the big-bellied men on the employment scheme care? Or Greg the supervisor, or Alma at the social security office. He imagines telling them, all of them, how he spoke up for art, and for a moment he is light-headed.

But they wouldn't have a clue what he was talking about. And they would never believe him anyway. And Mum doesn't even know who he is. He pictures her, sitting alone and gobsmacked and wasted. What was her life for? What was gained by it? There is nothing left of her except him. He breathes heavily. His face radiates heat. Feverishly, he fumbles in his pockets, crams his coins into the mechanism, pushes with his whole body against the glass.

The door gasps. He stumbles heart-pounding into a clean, modern vestibule. The air is calm. Wesley hunches, motionless, while the glass door sighs shut behind him, another in front swishes open. He steps tentatively into a bright, modern foyer with expensive-looking inlays in the panelling.

There is no one here.

In the silent, unmoving air, he takes the opportunity to sneer at the hidden lighting system, the perfection of the reflective walls. There is not even a desk. Where will he get his money back? Or even get information? A single sign – THIS WAY – points to a set of double-doors. Wesley throws his hands up.

The doors part, and he enters a huge space, rectangular, with a high ceiling. The walls are eggshell-white, the floor pale terrazzo. Everything is evenly lit. There is nothing on the walls, nor on the floor, or suspended from the ceiling. The doors close behind him, and they are as pale as the walls. Wesley moves cautiously forward. Turns to examine each surface. There is nothing. Or wait... no, it's an electrical junction box. Maybe that's something over there? He approaches obliquely, afraid to admit an interest. It is a fire-fighting point. He draws upright, indignant. At least there is no one to witness his humiliation. Not even security cameras, that he can see. In fact... there is nothing to focus on at all.

He turns, to go back to the foyer. Probably he overlooked a leaflet stand, or some information thing, headphones perhaps. He almost crashes into the closed doors. Looks around, for a motion sensor, a bell push, anything.

A very small notice: PLEASE USE DOORWAY OPPOSITE

Wesley huffs through his nose, strides the length of the room. Somebody, someone hidden, could be watching this farce. His skin shivers. But where could they hide? They can do anything these days. No surface is safe. The doors slide apart as he approaches, close behind him, and he finds himself in a similar room, identical except for a different orientation. His eyes widen. It too is completely empty.

'Hah!' he shouts.

The sound echoes. He slaps his hands against his hips, turns all around, wall after white wall. White ceiling. Pale floor. Nothing. He whirls to the centre of the room – why shouldn't he? The walls spin. An enormity swells inside him: all these rooms will be the same. There is nothing to see. He has spent his money, and there is nothing here. What a joke! But still he has some lingering hope: perhaps there are symbols to scan, Apps to download. You need a phone maybe, or a tablet, a computer. But there is no indication. Or perhaps the museum is not open yet? Perhaps he has come too soon. He laughs – an echoing falsetto. Imagine, at the police station: I paid my money, and the Museum of Future Art was not ready yet. What a skit they would make of him. He continues to slowly turn, spreading his arms like an aeroplane. He flies slowly along one wall, then along the other walls. He flies slowly to the door, into the next room.

It is exactly the same. White, enclosed, man-made. Yet it is infinite. Wesley glides effortlessly around, taking in the walls and the ceiling and the floor. He does not have to think, there is nothing to consider. He goes round and round, round and round. Passing into the next identical room, his eyes widen at the possibilities, the potential for beauty and understanding. Eventually, he turns for the next set of doors, flies through, and finds himself outside on the steps.

Two people are at the entrance door. A young couple, Spanish-looking maybe. The man has rectangular black-rimmed glasses, an ochre scarf inside a coat. The woman is squatting, picking coins off the ground. Four euros.

Wesley lets his arms fall. Feels his face specked by drizzle. He looks up to the lilac sky, to streetlights beginning to glow pink.

'Is this where the museum's to be?' the woman asks.

A moment passes before Wesley is sure that she spoke. The couple gaze at him, waiting. Wesley blinks.

'Probably,' he says.

He turns his head, to consider the pavement, the steps down to it. There is some great thing to be done here. The woman and man are watching, but even so, he takes a deep breath, raises his arms once more. He takes-off down the steps, and banks gracefully in the direction away from the bus station. That is enough. He burrows his hands down into the pockets of his coat, and tightens himself against the drizzle for the walk home.

# Sex for the Organism

I t was a shock to me when I realized that I was entombed in an organism. This blob of flesh could barely even sit up, let alone communicate or nourish itself. Worse, I discovered that this... this organism thing, which seemed to be my vehicle, would wear out and die, and apparently me with it, after what is called a "lifetime". So there was a definite urgency to the proceedings. Furthermore, I discovered that the organism would take a quarter of this "lifetime" just to mature: in other words, before I could use the thing properly.

So I had to get involved in this organic process. I insisted that the grossly immature organism carry out a little furtive research, then take certain hormonal supplements, so that it might reach a state of at least half-cocked maturity as soon as possible, and I might get some decent use out of it. But now that I have succeeded in boosting the thing into premature puberty, what does it want to do? It wants to reproduce.

You see, the organism has had an erection. Unfortunately, it is of the Y-chromosome variety, the testicle-bearing type, and all its spare capacity now tends towards matters sexual. And no, it has not considered what happens after sex, otherwise it would not want to indulge in the first place. Obviously I cannot allow it to reproduce itself. We would simply expend our existence rearing a new clutch (I observe this happening all around) and I would end-up having to run the organism myself, bringing myself down to its level, and there would be no time for my own affairs.

There were times when I did intercede in the running of the organism. It was I who forced the first premature words through its lips, which promptly sundered the nurture bonds of its parents, whom we have not seen since. The organism and I ended up "in care". Which was alright at first. But gradually the organism was made to feel disadvantaged by my involvement, and through psychiatric intervention and the use of sedative chemicals, the relationship between the organism and I regressed quite badly. It became plain that if we were ever to get out of "care", the organism would have to demonstrate an ability to conduct its existence in line with its fellows. In other words, I would need to be unheard. I took the necessary action, and we were discharged to a foster home, whereupon I immediately caused the organism to access some appropriate technology, and I effected a change of identity for it. Now I have it living independently and it is free of its medical history. But it had major doubts about this change of identity. How did I sway it? I promised it sex.

Unfortunately, the organism itself, being comparatively young, is barely capable of survival, let alone sex, despite the fact that I have found for it a less than onerous niche in its society. The niche I have selected for it is in a commercial enterprise where amalgams of industrial proteins and carbohydrates (known colloquially as burgers) are dispensed in return for tokens that they call money. The organism's involvement is to shout orders one direction and say the cost in the other. There are advantages to this niche. First: the brain is seldom required, so I have almost free access to that part of the organism with which I generally associate. Second: if, in an ecstasy of discovery, I cause the organism to discourse upon, say, a mathematical explanation of existence, its workmates tap their index fingers against their heads and laugh. And thirdly, but most importantly for the organism, there is a procession of egg-bearing organisms which have the facilities that mine requires for sex.

A particular egg-bearer that frequents the burger outlet has caught the organism's attention. I suspect this is due simply to the tightness of its coverings and the protrudence of its glands. Of course, I am immediately called on to help, because the organism is about as capable of introducing itself to the egg-bearer as it is of pirouetting atop a flagpole. It is afraid the egg-bearer will not take it seriously, and it has a point. In creating its new identity, I may have adopted for it an age that it is not physically ready for. To make matters worse, the organism is not a good judge of character: the inkings on the egg-bearer's epidermis, the multiple impalings of its lips and eyebrows – these are not good signs. But the organism will not listen. It begins to mope, my access to the brain plummets and something has to be done.

If there is one thing that became obvious during our confinements, it was the usefulness of chemicals in controlling the feelings of organisms. So, using our forged identity card, I cause the organism to effect the purchase of a distillate, and I coach the organism rigorously – when the particular egg-bearer asks for its beverage, the organism must say: "Would you like a special one of those?" A pre-prepared cup of distillate is then topped up with the requested beverage. The organism achieves that transfer, no problem. Then it loses its nerve. The plan was that it should take its break and join the particular egg-bearer. Eventually it works up the courage to do this, almost too late, because the particular egg-bearer and its friends are beginning to leave. However, it works out better, because the particular egg-bearer, which is giggling after consuming that beverage, waits behind, curious to see what is going on.

'What's your name?' it asks.

'Kevin,' says the organism.

'Kevin, you dirty dog. What's with the vodka?'

Of course, the organism gets immediately flustered and cannot speak the lines we have rehearsed. The egg-bearer laughs outright. There is nothing for it; I have to take over, to give the organism time to pull itself together. I propel from its mouth a simple discourse on the connection between sonic rhythms and the secretion of endorphins. Meanwhile, the organism gawps at the cleavage on display. The egg-bearer becomes somewhat agog at my discourse, so I switch subject to the relationship between ocean tides and mechanisms involved in preventing pregnancy. This induces a stunned expression. Until the egg-bearer looses a peal of laughter. Shaking its head, it opens its mobile phone and sets it on the table. I see that I must take an even lighter tone, so when the egg-bearer asks how long the organism has worked in that place, I opine on the current propensity to graduate the fourth dimension in relation to phenomena such as the rotation of the planet and its disposition to the moon. Each assertion causes paroxysms of mirth. In the end, the egg-bearer has to be slapped on the back to maintain adequate respiration. I use the intimacy to make an arrangement for the two organisms to meet again.

The evening has arrived when the organism is to meet the particular egg-bearer. With some difficulty, I talk the organism past the security at the door, and into the appointed institution. Inside, bursts of discrete sections of the visible spectrum provide disjointed illumination, causing impressions of confusion, novelty and constant reappraisal. This makes banal organisms seem more interesting to one another, which is suitable to our needs.

We find the particular egg-bearer.

'Kevin!' it yells. 'You're a fuckin star on YouTube. I put your speech on it and you got ten thousand…No…? Oh I forget! Never mind! Lookit, Tammy here… It is Tammy, isn't it? Tammy even came from… where was it? Ah never mind! She barely even speaks English! Here!'

It bawls across at its other friends.

'Meet Kevin. He's a fuckin scream! Go on, ask him something! He's fuckin mad!'

I gather that this is not the first establishment they have been in this evening. Thankfully, oral communication is almost impossible due to the amplitude of the entertainment, so the organism is able to gawp freely with its mouth open. Meanwhile, I hope that the YouTube audience did not include our former care professionals, otherwise the organism may need another identity change. However, this will not affect our current mission. I instigate a dance with the particular egg-bearer, but I fear the jagged prancing of the organism contrasts unfavourably with the harmonious if rather overt routines of the particular egg-bearer, which draw appreciative glances from other bearers of testicles.

Back at the drinks table, the one named Tammy speaks in a thick exotic accent.

'Bottom is up,' it says.

They all giggle at it and empty receptacles into their throats. I notice that the Tammy brings distilled drinks for all the egg-bearers but purely aqueous solutions for my organism. Gradually, the particular egg-bearer becomes less likely to discriminate against my organism's physical shortcomings, because it can barely focus. My organism also is beginning to wobble; it has never experienced alcohol before and even the little it receives is having a significant effect.

Commercial activities cease. The egg-bearers rush about, yelling and looking for their coverings. The organism's vision swirls from interior views to unsteady street scenes, with images of railings, arboreal growths, buttons, zips and short vegetation. Surprisingly, we are presented with a panoramic view of half the universe, eclipsed almost immediately by the Tammy egg-bearer unburdened of its coverings. Its face is calm and blank. The organism is lying uncovered on grass, and the Tammy egg-bearer gets on top of it. Words are spoken, but the organism does not bother processing them because, due to inexperience in these matters, it has proceeded immediately to climactic overload. Nevertheless, there continues a period of intense shaking. The Tammy egg-bearer takes on an attitude of quizzical pain. Its mammary glands achieve harmonic motion, and the shaking intensifies until the universe is a kaleidoscope of bright-jagged lines. There is a wrenched gasp, and the organism is buried under skin and hair.

That is the problem with alcohol, one cannot control what will happen. At least the organism has had its way, and as long as it does not take-up this activity on an ongoing basis, causing fertilization and all that entails, then this is not a bad outcome. I have kept my promise and perhaps…

But the Tammy egg-bearer has reared up. It begins to speak.

'We saw your organism on internet,' it says, 'and we recognise what it is. I will provide my organism for its coital needs, and those may cancel needs of mine. In such closed system we can so use our facilities maximum.'

Is this a joke? But then, like a first view of the firmament, I understand. It is speaking to me. I cannot at once comprehend the possibilities, but one thing is clear: I am not alone. I direct the fuzzy gaze of my organism to make certain I miss no optical data. What control it has of the Tammy! But there is no time to lose: we are both on this insufferable lifespan.

The organisms are kneeling, facing each other. It speaks: 'We will live the organisms together, maybe our place. Do yours have effects it need to bring?'

With some effort I bring mine to its feet, intending that it will fetch its few artefacts.

'Whash ish your addresh?' I make it say.

'Not this country. We will wait here to bring you. But put on its clothes before you run it.'

So. Another organism for sex is one thing, but the possibility of a fellow with which to discuss… well, I am affected. An organism's lifespan is not a long duration, but there is something more to it now. I dress the organism, turn it around, and run it for all it is worth.

# Binman

Xavier worked as a binman from 6am to 4pm, six days a week. He used to work for the city council, but they had outsourced the service and now he worked for a private contractor. The longer hours and faster pace of work did not bother him. He jumped on and off the truck all day, pulling the wheelie-bins, attaching them to the lifting mechanism, and leaving back the emptied bins exactly where they came from. Sacks were more difficult. Over the years, his right shoulder had become strained from swinging sacks of refuse up into the compactor, and lately he had to use only his left arm. Now the left shoulder too was becoming strained.

At lunchtime every day he ate four sandwiches. He brought these with him in a blue tin that had pictures of biscuits on it. The sandwiches were made by Astrid, the woman he shared a flat with. They were of brown sliced bread spread thinly with margarine. Two sandwiches had a filling of easy-single cheese, one had tomato with salt, and the last had strawberry jam. This was the combination he and Astrid had settled on.

When work finished, Xavier cycled back to his flat, which was in a large block on the edge of the city. Astrid would have left a plate of dinner ready to be heated. She worked in a hospice, and got home at twenty-five minutes past six, because it was nearly an hour's walk away and she did not have a bicycle.

Today was Tuesday, so dinner was pasta with cubes of pork in a sauce made from a tin of tomatoes. Xavier put it the microwave to heat it up. Usually he would take fifteen minutes to eat the food. Then he would have a rest for twenty minutes, or if something needed fixing in the flat he would do that. This left fifteen minutes to walk to the Community Centre. He walked to the Centre because there was no safe place to lock a bicycle. Two bicycles belonging to Xavier had been taken before now and it had been a nuisance. At the Community Centre, he would get a plastic refuse sack and either the litter-picker or a scoop for dog shits, and begin whichever route had been assigned by the Local Environment Committee for that day. This was voluntary work. Xavier could not remember when he had started doing this, nor could any of the present committee. But today Xavier realised that he could not go. He had managed to get through the day, but after cycling home and climbing the stairs, a swelling in his crotch had become so large, the pain so intense, Xavier realized that it was physically impossible for him to go to the Community Centre. Nevertheless, he ate his dinner.

After finishing the plateful of food, he was still hungry. Before this week, this had never happened before, and now for the third day in a row he was hungry after finishing his dinner. On Sunday, the day he first noticed a swelling of his testicles, he had said to Astrid: 'I am still hungry.' Astrid had looked at him, and he had looked at her. He had not said anything about his testicles, nor was the swelling much noticeable then, but she said: 'Is there anything unusual about your testicles?'

'Yes. They are a bit swollen.'

She nodded. 'I will make some more food.'

She had made porridge, because all the other food she bought was for some particular meal of the week.

The day after that, Monday, he had been hungry again after dinner, but because Astrid was not there, and it was her job to make something extra, he had gone to the Community Centre without extra food, and had come back weak and ravenous. Before he went to his bed, Astrid had seen him looking at the bulge in his crotch.

'Is the swelling getting more?' she had asked.

'Yes,' he said.

'Is there much pain?'

'Not too much.'

She had made some extra food without being asked. Then she rummaged in her drawer until she found a card.

'I am going to make a phone call,' she said, and she left the flat.

After she came back, she did not say anything. He had thought that maybe a doctor might come. But no doctor came. They each took off their over-clothes and went to their beds.

This morning, the swelling beneath his underpants had been very much bigger, and when he had first tried to get up from bed, the pain had made him gasp. He did not take off the underpants at that point, because it was Tuesday, not Sunday. When he raised his eyes, he met the level gaze of Astrid. Her face did not flicker. She looked at the swelling and made a nod. Xavier had wondered if she would do anything, because it was her job to look after his sicknesses, but Astrid went as normal to the cooking part of the room, to make porridge and prepare his sandwiches, while he shaved his chin and cheeks and neck. He came slowly to the table, wide-legged, and sat with his knees as far apart as they could go. Astrid opened two sachets, emptied white powder onto his porridge and mixed it in. She had never done this before.

'You can still go to work for today,' she said.

After eating the food, he had become curiously numb. He had remained numb all the way to work on the bicycle, and for most of the day. But by the time he returned to the flat in the evening the numbness had worn off.

Now, because he was still hungry after eating the dinner, and was unable to go litter picking, Xavier did not know what to do. Astrid would not be back for an hour and forty-five minutes, to make some more food. While he was sitting at the table thinking about this, he was riven with such a pain that he fell onto the floor. It felt like someone had their fingers in his genitals and was twisting them apart. Lying on the floor, he observed the crotch of his workpants, which was ballooned out because of the swelling. It was impossible to touch his crotch, even to relieve the pressure by loosening his clothing. Another shaft of pain transfixed him. And again. His body arched. Yet within the agony, his overwhelming feeling was of hunger.

He struggled to his feet, legs wide apart, and made little twisty jumps toward the fridge. It was not his job to use the fridge. He had rarely seen inside it, unless he happened to pass while Astrid was putting in or taking out food, but he opened it now. On the top shelf were the small parcels of meat or fish, and larger parcels of vegetables that Astrid bought on Saturday for dinners during the week. But on the shelf below were a number of unusual items: packets of solid orange cheese, and some plastic cartons of cream. He had never before seen or tasted items like these. He opened a carton of cream and drank from it. He held the carton in one hand, and lifted one of the blocks of cheese with the other, biting deeply into it. A spasm of pain caused him to spill some of the cream, but he held on, and putting the carton to his mouth again, he poured and sucked and glugged. He crammed more cheese into his mouth, gulped it down in lumps. He tried to finish the cream without spilling any more. At the next spasm he almost lost his sight.

The door of the flat opened and Astrid came in. She was home early, the first time Xavier had ever known this to happen. She was carrying a large brown box that contained cartons of cream, blocks of cheese and large tins of a type that he had not seen before. It was Xavier's job to carry heavy shopping, but today was Tuesday and anyway he was on the floor. She did not seem surprised to see him on the floor. Her mouth was small as always. It never moved except to speak or eat. There were no creases on her face, although the way she moved gave the impression that she was older than Xavier. He had never touched her skin that he could remember.

She said: 'You must get on your bed now.'

Xavier was impressed by her knowledge. Sometimes she knew things he didn't think she could know. She laid the box on the table, and went out to the hall. He tried three times to get on the bed. Each time a tearing pain put him back on the floor. Astrid came back with four trays of eggs. She put these on the table, then came and lifted him under the armpits as he was trying to get up. His next spasm caught her by surprise and both of them fell over. They got up and Astrid rushed him toward the bed. His crotch was squeezed during this effort, and his vision became so bright he could not see.

He opened his eyes and Astrid was tying his arm with a strap to the top corner of the bed. His other arm was already tied to a strap on the other side. His legs were tied apart at the bottom.

'What are you doing?' he said. His voice was thick and slurred.

'You must not damage yourself.'

'But…'

His body arced. He jerked and twisted on the bed. The pain seemed oddly dull, but it filled the room. The straps held him.

Astrid looked down at him. 'I have tried all the hospitals. They have no bed.'

'Doctor.'

'No need for doctor. I know what to do.'

'Food,' Xavier said.

Astrid went to the table. She lifted a large bowl, and a short wide spoon that was already coated in yellowish sludge. She brought these over, buffed-up the pillow beneath Xavier's head, and spooned a thick yellowy mixture into Xavier's mouth. He slurped until all the mixture was used up.

'More.'

Astrid worked steadily. She cracked half a tray of eggs into the bowl, poured in a carton of cream, four cupped handfuls of grated cheese and half a tinful of powder. She mixed this around with a wire beater, and came over and spooned the mixture into his mouth. She made another bowlful, and spooned this too into his mouth until Xavier was satisfied. He closed his eyes.

Voices. Foreign accents. Xavier opened his eyes. Two men sat facing the bed. Another man paced around, even though there was a chair for him that Xavier had not seen before. All three wore suits, of different styles and fabrics, expensive-looking, but creased and dishevelled, as if they had slept in them overnight. The pacing man was slightly stooped, blondish, pale skinned, and the man seated nearest the door was olive skinned, thin, with close-cropped bristly-grey hair. The man in the centre chair sat in a slumped manner, hands together making his fingers splay. His hair was dark and oily, his lips wet. This man said wearily: 'So he is awake, one more time at least.'

The man who was pacing stopped, looked briefly into Xavier's eyes.

'So he is. Yes.'

The dark-haired one leaned forward. 'Tell me, do you know who you are?'

'Xavier...'

'What does it matter, at this stage, if he knows or not?'

'It is interesting,' the dark-haired one said. He leaned again. 'Tell me, do you have any idea what is happening to you?'

'No.'

'Do you remember this from before?'

'No.'

'You don't know what this is?' He motioned toward Xavier's crotch.

Xavier raised his head and looked down along his body. He was naked, and his limbs had become thin, the colour of old cabbage stalks. In his crotch two monstrous masses slowly writhed. They were covered in thin pink skin, with long purple veins like rivers in an atlas.

A loud knocking at the door brought Xavier again to his senses. Astrid stood up from behind the men and opened the door. A young man in a pressed blue suit breezed confidently in, but halted abruptly, as if cobwebs or some such thing had touched his face. He put a hand over his nose.

'Phew!' he said. 'Good God...'

He saw Xavier, and his eyes widened.

'What on earth...?'

The dark-haired man said: 'You are the health authority administrator?'

'Yes, but... what the hell's this?'

'You will not admit him to a hospital?'

'I got your note, but... look, first of all we've no available bed. And second, I just don't believe what you say. What the hell's going on?'

'I take it you do not read the literature.'

'What literature?'

'What literature? Biology. Sociology. Chemistry. Genetics. The writings of science, the workings of the world.'

'Evolution,' the olive-skinned man said.

'Look, first I need to know what's caused this... this..." The health authority man flapped his hand at Xavier.

'Caused? You caused. I caused. It is a consequence, an inevitability one might say, of progress.'

'Food...' Xavier gasped.

'So you guys didn't do this to him?'

The three men laughed.

'No, my friend. The first time, thirty-three years ago, we saw the signs and came to watch. Now we give a little help. I contribute finance, this man gives medicines, and this man makes some surgical interventions. To get over the hump, as it were. These are our gifts.'

'But I've never seen anything like this. How did you find him?'

The dark-haired man shrugged, moistened his lips. His tie was loosened, and his beard had grown since Xavier first saw him.

'The first time? I myself had been mapping patterns of population, constructing a model of evolution, which led me to cities such as this one. After that it was medical records.'

The olive-skinned man leaned in. 'We followed behavioural anomalies.'

'And we'd been monitoring the effect of pollutants, following effects up the food chain.'

'Pollutants? Anomalies? I don't see–'

'Come Sir! Surely you do not think you could isolate mankind from the natural world, saturate the biosphere–'

'Interfere with evolution.'

'Foster eusocial behaviour, as if we were bees?'

'Stop!' the young man said. 'Who are you guys?'

'We are simply men of science. We observe the development of our species. We are here to observe.'

'I don't know what you're talking about!'

The man who had been pacing snorted, tossed his head. The man with the dark hair sighed: 'It is as though there is a bright pulsing light, and yet nobody sees it. It is even in the papers from time to time. Pollutants. Population. Monoculture. The annihilation of diversity, the overwhelming of our niche. What did you expect? Our species is merely reacting, adapting to a steady state–'

'Food… please…'

'But what's the connection with…with…?'

The olive-skinned man laughed harshly. 'The connection? In the rivers of this country, fish have been growing eggs in their testes for over thirty-five years now. That was caused back then by just one minor synthetic chemical in a sunscreen product. Today the biosphere has over sixty thousand such chemicals.'

'No no, it is population. Population density is one of the main–'

'Gentlemen, please!' the oily haired man held up his hands. 'We may debate this later, but this may be the last feeding, and look – after all he's been through, he becomes conscious at the end. Isn't that amazing?'

Astrid spooned the mixture into Xavier's mouth. He glugged it greedily. After a time, he began to tremble. His sight faded.

He sensed himself fighting to breathe, thrashing his limbs around. He was confined in some thin covering, some membrane, that he had to tear his way through, to burst and rip his way into pure fresh air. He whooped-in lungfuls. Wiped his eyes. Banged his frail limbs against something, someone, and fell off a bed. There were three men sitting on chairs, and a woman. He raised himself weakly from the floor. Felt light and young and hungry, wiped slime from his face and limbs. A flaccid cord extended from his belly button, up onto the bed.

One of the men stood up. He had olive-coloured skin.

'Excuse me,' the man said.

He put a clamp on the cord right at the belly button, cut the cord with a clean steel scissors, and gathered the remainder onto the bed. A collapsed and emptied body was strapped to the bed, unmoving, its jaw stretched open. Two limp tubes were heaped in its crotch, and on the other side of the bed, curled and panting, was a slim, wet, young-looking person, who was strangely familiar.

# The God of Thin Women

My wife has me completely confused. I had always imagined that calories ingested, minus calories used, equals weight gained. It seems simple. Yet for months on end she has convinced me that certain foods, under certain conditions, are not fattening. For example, if you really like a certain food, it is not fattening. Food is also not fattening if it might otherwise be thrown out, if no-one sees you eat it, if you eat it in small pieces or off a small plate, if you don't *normally* like it, haven't eaten it for ages, if you decline it at first before accepting a piece, if you eat it in the dark, don't sit down while you're eating it, have had some drinks, just met a friend, or have simply had a bad day. She convinces me of this. It is not for discussion. I dare not mention past situations. In fact, I dare not say anything. To comment is to be against her. To even have an opinion is to be politically incorrect.

There are times I would rather not have an opinion. For instance, when the Faculty Ball is imminent and she is fitting on an evening dress. At the first mention of this, I slink to the furthest recess. You see, I have become used once again to the happy eating, but the tide of our lives can so easily change. I cannot go through all that again, and it might take only an evening dress to tip the balance.

'Norrr-mannn!' she calls. Do the windows shake, or is it my imagination? 'Come here, will you?'

There is no option but to trudge to the bedroom. I try to avert my eyes, but there, in an evening dress, my wife wobbles a tiny pouch of flab where only her svelte tummy ought to be. She looks at me keenly, and asks if she has put on weight.

'No,' I say, 'you have not put on weight.'

But I know already it is hopeless.

'Now,' she says, pressing her teeth together. 'Let's see if we can get you into your tuxedo.'

It is no surprise when she announces the diet. She never actually says that I too am on the diet, but, as usual, I will be collateral damage. What she eats, I eat. For a start, she cooks it, but in any case it seems unacceptable that I should eat, say, a meat pie, or banoffi tart, while she is having hummus.

'It'll be good for you,' she says.

Previous diets have left me distended with flatulence. Or glowing with carotene. Or reeking of asparagus and garlic. After a diet sweeps through the house, I usually collapse, listless and dejected, but she eats wolfishly. At that stage she has proved something, and she has the moral right to eat what she likes. Or else the diet has made her a victim, and she comforts herself, or reasserts herself, by eating.

I have survived many diets, but they have taken their toll, and like a child turning on and off the TV; on-off, on-off, on-off; one time I am afraid the switch will not work. Furthermore, there is something ominous about this diet. For a start, a number of women are involved. They are having meetings. A variety of tactics are under consideration. A holistic approach, she says grimly, her eyes shining with martyrdom. They mean to try everything at once – a multi-pronged assault on body, mind, and spirit. I hear mention of beta-drugs, untested Amazonian distillates. They spend hours discussing supplements, tonics, elixirs. The herb shelf overflows with potions unnameable. The hall piles-up with self-help books and the latest exercise wear. The house echoes with the crack of celery, the fridge fills with grapefruits and watercress. She finds a source of lettuce that is genetically engineered to provide no nourishment at all. On the TV screen, the gastronomically undead perform exercises in bony synchrony. In the bathroom there are gravity-negating face-creams, tinctures for every sign of the zodiac, unguents purporting to turn back time. There is mention of Chinese acupuncturists, pygmy herbalists, Bantu witchdoctors. Does she have any idea of the forces she may unleash? If even one tenth of that stuff actually works, the results would be incalculable, and what would happen, in the name of God, if all of it worked at once?

I lock myself in the study and the walls hum with dread.

The day approaches. There is always a start day. Like a gun goes off and she starts fasting. I cannot bear it. I betake myself to an academic symposium in a distant city. But even there, I cannot sleep for fear of what may be happening back in the house. She will be having herself hypnotised. The air will reek of heated oils. She will purge her insides, exfoliate her outsides and starve the bits in between. She will deck the place with amulets, charms and crystals. Nothing will be left to chance.

After a week away, I am bug-eyed with trepidation and lack of sleep, almost delirious, and I am not even on the diet yet.

When I return late at night, there is indeed a fug of occult smoulderings about the place. I creep into the darkened kitchen, slavering for a snack, but everything I find would take more energy to consume than it would provide in nourishment. A biscuit falls to dust in my mouth. I wash it down with something called Low-H2O. The side-table is like a shrine: there are scented candles, vases of flowers, books strategically open. In the centre is some sort of statue – probably the God of Thin Women – of such attenuated physique that I am not sure if I can see it at all. I plod hopelessly up the stairs. In the near-dark bedroom, I have not the bravery to look at her – heaven knows what she has done to herself. She barely murmurs goodnight, and it seems that her voice is reduced almost to nothing. I seep into the bed, delusional with hunger and anxiety.

This time she has gone too far.

When I wake in the morning, my worst fears are exceeded – she has faded away completely. Where her head should be on the pillow, there is only an empty hollow.

Why didn't I stop her? But who could stop her, I'd like to know. Trouble and all as she has always been, tears run down my cheeks.

*What on earth's the matter with you?*

What? Is that her voice? – coming from somewhere by the pillow.

*What are you staring at?*

'But... my dear... you're...'

*Well, stop babbling and give me a hand. I can't seem to move a muscle.*

'But my dear, you haven't got... you haven't got a...'

*Here, pull this duvet off me – it weighs a ton.*

I look down: there is no duvet, only a sheet. I peel it back tentatively, in case some tiny sliver of her has survived.

*Ooh... that's better! I feel light as a feather.*

I tumble from the bed, so confused that I put my trousers on over the pyjamas. Is it any wonder? I hasten around to her side, to let up the blinds. The sun is dazzling.

'Em... my dear, surely you should rest today. Take it easy. This dieting… it's taken too much out of you…'

*Oh don't talk rubbish.* Her voice comes from near the window. *I might go for a walk along the river. Or a run.*

I stare hard and long, and eventually, perhaps, I perceive just the barest line of interference in the dust motes.

*Tell me* – her voice is now behind me – *do you notice any difference yet?*

I whirl. There is only the merest diffraction of light to show where she might be, heading toward the long mirror. I throw myself in front of it.

'Don't look at that! You've definitely overdone it this time.'

*Nonsense. Never felt better in my life. But now, what should I wear…'*

What will happen if she tries to lift even the skimpiest undies?

'Why…um…why don't you have some breakfast first? You must be famished.'

*Not the slightest. Couldn't care if I never ate again. And the same with clothes. Why bother with them? We should liberate ourselves. I'm going to go naked, I don't care.*

A carefree sigh heads for the bedroom door.

'No!' I cry, and dive in front of her. Whatever chance she has in the bedroom, how can she survive in the kitchen? Suppose a draught of air catches her, dashes her against the saucepan rack? Or she is sucked up by the extractor? I have to keep her confined, go and get help. But to whom do you go in a situation like this?

*Come on, Norman. It's a beautiful day and I want to make the most of it.*

She is already beyond the door, going down the stairs. She must have stepped over the hinge. I bumble after her.

'Wait! Wait, lovey! What's the hurry?'

*I'm supposed to go for a jog this morning, I told you last night.*

'What? You can't!'

*I beg your pardon? I'm going for a jog and that's that. And I don't want anyone spoiling this beautiful morning, so if you can't be cheerful, kindly keep quiet.*

'Yes, but…'

I have to protect her. Her running gear is in the utility room, and if she tries to even lift it, the truth may be too much for her.

'Listen, my dear. It's just not the thing, not today. Wouldn't you feel like coming back to bed?'

*Why Norman! I was wondering when you'd take more interest. Yes, certainly, let's go back to bed. But first I need a little run, ok?*

'No, no. You can't put on that gear.'

*You're right. Why should I wear that stuff?*

'What... what do you mean?'

*I'll go naked. Most natural thing in the world. You have to love your body to treat it properly.*

Her voice is heading for the back door.

'Wait! I'll come with you.'

*That's the stuff! Get them off you then!*

'What?'

*Well, get your clothes off. If I can do it, you can do it.*

'I'm not going out there naked!'

*Oh, just a little jog? You've never been for a jog in your life.*

'I'm not going naked! On the public road?'

*Oh come on! I'll tell you what, just around the lawn then. It'll be a start.*

'But what if there's a breeze? I'm worried that... you know...'

*It's not a bit cold.*

'That's not what I'm worried about.'

*Look, if you're not coming, I'll go along the road.*

'Ok, ok, once around the lawn then. But you have to stick right beside me, ok? And if there's any wind we come straight back in.'

I take off my trousers and pyjamas.

'Now we stick together. Right? And just once around.'

*Oh all right then. It'll be worth it to see you doing something energetic.*

We set off around the tiny lawn, me a wobbling ivory-coloured pudding, she a blade-cut in the air.

There is a sky-splitting screech. Shrieks of laughter. Arriving at the back door is my wife, with her friends, Jenny and Anne. They are in day-glow jogging lycra, red-faced and perspiring after an early morning run. Jenny and Anne cover their mouths, collide in their giggling hurry to get inside. My wife stares, drop-jawed.

'Norman, what on *earth* are you doing?'

# Don't Start Reading This Story

Ok, now do not stop reading this story – you see? – it's got Natasha Dobronovski in it. Or rather, *I've* got Natasha Dobronovski in it. She really is something to look at. She's not a character in this story – she's not even in the story – but she's a vital element nonetheless. You see, Natasha is my hostage. How I got her is another matter, but now I have her and that's all you need to know. And now that you're reading this, YOU are responsible for her safety, because if you stop reading this story – if you go anywhere before this is finished – then I am going to do some absolutely unspeakable things to Natasha, ok? You can pretend this isn't happening, that you're only reading a piece of fiction, but I tell you now, it works like this: you keep reading, nothing happens to Natasha; you stop reading, Natasha gets the treatment, and everyone will know its your fault. That's the deal. It's a straightforward hostage situation.

Now get this. This story is specifically for you. Yeah, you. If a story is properly written, everyone gets a different meaning from it, and what you're reading here cannot be understood in the same way by anyone else. At this stage, I almost wish you could see Natasha; the *lustre* of her hair… how does she get it to look like that? One would like to reach out and… no, that's stupid. We're not going to have descriptions of Natasha. You'd probably get more interested in the hostage than the story – and that's not what I want. Let's just say I went to a lot of trouble to select her. I needed someone of universal appeal, of irreproachable innocence, so that if you let me down the repercussions will be chokingly awful. So now, read or leave, it's your call.

I see you are still here. But you don't like this *involved* writing, do you?

You prefer that domesticated stuff – which is exactly why I have taken this action. How the hell is a writer ever to get a real live story read when people lap-up that safe crap that lies face-down on the page? It doesn't involve real characters, let alone the reader. I mean *involve*, not *interest*. I know, I know – you like that stuff, because you can keep it at a remove from yourself. You feel safe reading it; you can cuddle up with it and bring it to your bed or your fireside chair; it won't threaten you. But how can a writer deal with life when you only read safe fantasy? And it's all fantasy. Don't give me that genre bullshit – it's all the same. What I want is for you to actually *feel* this story here. I want you to be *in* it.

Now, of course, you've forgotten about Natasha because you're thinking about yourself. Yeah, to hell with poor trembling Natasha. But if you only read what you want to read, then the reader *is* the writer. It's the reader who gets things published. Sure, you don't actually press the keys – any old donkey can do that. And so with all the other donkeys on the chain: the submission editor, the publisher; all these are about as important as the truck driver or the sales assistant in deciding what you read. What you are doing is making writers extinct, and what I am doing is fighting back. And that's why I've brought Natasha into it.

Of course that's not her real name. I'm not totally stupid — you could be reading this and dialing Rapid Response; that is, if you appreciated the gravity of the situation. Denial is the first reaction to crisis, because denial is so easy. Anyway, calling her something exotic like Natasha was more certain to get your attention. The more unlikely a thing is, the more likely you are to believe it. That's fiction. But watch that word: fiction means it's made up, it doesn't mean it isn't real. In fact, this whole thing wasn't real until I made it up, but it's real now. Which is unfortunate for Natasha, who is certainly real although her name isn't. Oh, if you could only see her now: the first thing she does when she's upset is get out her make-up mirror. Isn't that touching, eh?

Maybe at this point you grasp a straw of hope: this writer's a nutter! – you say. Or some convenient James-Bond-style baddie with a crippled body in an electrified wheelchair. You wish. You prefer to think in fantastical terms so you can absolve yourself from what I will probably have to do. I say probably, because I see you getting restive again for your cosy wrap-around sentences and your just-new-enough topics – oh, and a nice convenient genre to carry them in. You don't like being actually responsible for a character, let alone a real person. You think you've never been in that position – but that's just what I'm telling you; you've always been responsible, just no one has pointed it out. What happens in literature is your fault!

Very well then, remote from your sight, let me lead our helpless Natasha to where we have easy access to my little workshops. What will it be? Humiliation? Depravity? Mutilation maybe? It's no matter to me. Just don't expect to be regaled with details. Not while it's happening at least. I'll be damned if I'll reward your callousness by documenting Natasha's screams, or with any lurid detail of what drops to the floor. Natasha's shrieks can echo in your subconscious as you go about your comfortable business. Just don't blame me afterwards. For me it's survival; I've set out my stall and I'll do what I must.

Now, perhaps, you think you might simply destroy this story, hoping that no one will know. It isn't so easy these days to bonfire inconvenient literature. Do you think I'd risk everything on a single shaft? No way. The printed page is merely a fructification, the mycelia are already pervasive. It's too late for fire.

But – you argue – everyone is reading this same story. Don't bank on it. It's an easy business to change each print, let alone each soft copy. Nothing wholesale needed; a word here, a nuance there. Even when the story is exactly the same, it can easily be structured so that different people get a totally different impression from the same story – remember English class in school? 'Ha!' – you think – 'this writer is trying to frighten me with writing, but we are in different places, perhaps different times. I cannot be got at.' Sure. And yet you're not a bit surprised when you learn in history that a tome of writing incited one entire nation to incinerate another. Or that a little fictional "report" was used to start a war in a certain oil rich country? Or a compendium of musings written centuries after the fact sets an entire religion against all others for over a thousand years? Wakey wakey! Writing does the business and it's a dangerous business. You don't know what's real and what's not, and it doesn't matter, because often it isn't real in the first place, but then writing makes it real, and now YOU are the writer of what will happen to our mutual dependant, Natasha. Not the details of course; I will look after the details. They may not fit your usual genre (they certainly won't fit Natasha's).

And yes, yes, I know this is wrong. And if I'm caught, I'll be… But damn it, what choice do I have? You, and others like you, have imprisoned *me* in ways you cannot imagine.

Of course now you wonder: how will this "writer" know if I have read the story? Don't worry, I will know. A sign will not light up, it will simply become obvious. Perhaps you hope that by the time it becomes obvious,

Natasha will be sitting with her friends in a café, telling them about a most bizarre and unsettling experience she has just been through: a place she didn't know, forces she couldn't understand – her chastity, innocence, even her life itself, fragile in her shivering body – and then she was simply set free. Now I ask you: is that likely to happen? A desperate writer risks everything; plans a meticulous abduction, launches an integrated series of fictions and events, bares an impoverished, writerly soul to the winds of contempt… and you think you may get away without the pivotal consequence? You've got a bad case of happy-ending syndrome. If I don't get my way, Natasha will not come blinking and innocent into the light. No chance.

So now, your last hope: get the law to rescue Natasha. An incident room, perhaps in a disused publishing house. With computers. Whiteboards. Clippings pinned to the walls. A team of veteran sleuths. They could collate rejection slips, geographically reference them against recent SAEs, fractally analyse plots and character developments. They could cross-reference themes, identify a hard core of starving would-be authors who might – just might – be desperate enough to try something like this. I bet you already have preconceived ideas about the gender and social level of this writer, and that may be a help. Or maybe not. You would of course winnow out academics, poets, librarians and teachers, rank inversely for age, filter the whole mess by sexual orientation, and in a breathless finale, some nicotine-fingered coffee-sodden marriage-outcast will have an absolutely perpendicular insight that sets a car-load of mixed-gender sharp-dressed athletic gumshoes pounding up a bare stairway to shoulder aside a paint-chipped door, to surprise a one-finger typer who immediately tries to swallow the vital page that describes where beautiful, helpless Natasha is incarcerated. They leap upon the mad writer, but despite their heroic efforts the masticated lump of pure literature is gulped past a tourniquet of tightening fingers, and the bloodshot eyes of the evil keyboard-basher blaze with triumph. Then, and only then, do YOU breeze in, after your dash from the incident room, you dive to the computer, and with clenched fist you hit PRINT! And Natasha is saved.

Such absolute crap.

That isn't going to happen and you really have to stop reading stuff like that. The shimmering amalgam of life and fiction does not work that way. Only a writer can know the utter vacuum in which Natasha is entombed, and it is simply not possible to get her out without me doing it. There she would live, and there she would die. But I will concede one thing: you are still here. And that least of things is in obverse the most of things. It's lucky for you that this is a short story. It could have been an airport book. Or a trilogy. But 2000 words is the limit of our involvement today, and as such you have fulfilled the contract. You have read the story. So now, just for you, here is this:

*"Natasha stumbles into daylight, holds out her hands to steady herself in the unfamiliar street. Her goosebumps fade; the warm smells of fast-food and traffic condense on her clothes and skin. She totters to the kerb, amazed that she still has her handbag, and raises her arm for a taxi."*

So, this has been a reconnaissance mission, in a way. I have seen what is needed to get a story read, and you have glimpsed the squirming mess on the underside of literature, and your part in it.

Now Natasha is having coffee with her friends again. She feels she should be pressing charges, but she cannot say against whom, or for what. Frankly, I don't think her friends believe what she says happened to her. They are sympathetic, but they just want things to be normal. They don't want the world to *be* like that, and so they think – ha! – that she is making it up. But if I am to get you to read another story, I'm afraid I will be visiting her again, or someone like her, and things will have to happen.

I'll be in touch.

# Acknowledgements

To Catherine Phil McCarthy, who in the beginning was a lone voice of encouragement and advice. To Dominic Taylor and everyone in the Limerick Writers Centre for promoting writing in Limerick, and for publishing this book. To the Munster Literature Centre, particularly Pat Cotter, for championing short stories, and organizing a fabulous festival year after year. To John Givens, Claire Keegan and Mike McCormack, whose workshops or seminars honed my appreciation of story and craft. To past and present members of writers groups: Andy Hamilton, Fiona Clarke-Echlin, Kevin Liston, Margaret Cahill, Patrick O'Flaherty, Roger Marsh, John Carew, Sarah Moore-Fitzgerald, Clare Dollard, Mark Lloyd, and especially Caroline Graham. To all those who organise literary journals, competitions and festivals, for creating space for writers to inhabit, especially Southword, Revival, Crannog, The Penny Dreadful, Irish Independent, Irish Times, China Writers Association, Pure Slush (Australia), RTE, and Hennessy New Irish Writing 2005-2015 for first publishing stories in this collection.

To my mother Sheelagh and late father Pat, who gave us all freedom to grow, and to my brothers and sisters, whom it is hard not to tell stories about.

Finally, to Joan and our two young men, Padraig and Art, who all together are the essence of life.

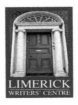

## ABOUT THE LIMERICK WRITERS' CENTRE

The Limerick Writers' Centre, based at 12 Barrington Street in Limerick City, is a non-profit organisation established in 2008 and is one of the most active literary organisations in the country. We endeavour to bring ideas about books, literature and writing to as wide an audience as possible, and especially to people who do not feel comfortable in the more traditional arts/literature venues and settings.

At the Centre we share a belief that writing and publishing should be made both available and accessible to all; we encourage everyone to engage actively with the city's literary community. We actively encourage all writers and aspiring writers, including those who write for pleasure, for poetic expression, for healing, for personal growth, for insight or just to inform.

Over the years, we have produced a broad range of writing, including poetry, history, memoir and general prose. Through our readings, workshops and writer groups, our aim is to spread a consciousness of literature. Through public performances we bring together groups of people who value literature, and we provide them with a space for expression.

We are, importantly, also dedicated to publishing short run, high quality produced titles that are accessible to readers.

At our monthly public reading the 'On the Nail' Literary Gathering, we provide an opportunity for those writers to read their work in public and get valuable feedback.

The centre can be contacted through its website:
www.limerickwriterscentre.com

Lightning Source UK Ltd.
Milton Keynes UK
UKHW010647200721
387465UK00002B/606